The Space Elevator

Bradley C. Edwards, Ph.D.
Eric A. Westling

The Space Elevator

Published by BC Edwards
7715 Twin Hills
Houston, TX 77071
eawestling@juno.com

ISBN 0-9746517-1-0

Printed in the United States of America

Table of contents

Preface

A wise man once said, "If a great scientist says that something can't possibly be done, then sooner or later it will be done." One-day, a few years ago, I read a statement that the space elevator couldn't be done, and I set out to find out why. From there things got *very* interesting and resulted in a research proposal being submitted to NASA. The proposal was funded and resulted in, first a six-month study and then a two year study. The core of this manuscript started out as the technical report from the six-month investigation I conducted for NASA under the NASA Institute for Advanced Concepts (NIAC) program.

That study had the same simple title as this manuscript, *The Space Elevator*. The study itself was far from simple however. The object was to investigate all aspects of the construction and operation of a space elevator, a concept that up until this time had been confined, first to speculation in a few technical articles and then to the realm of science fiction. Our original six-month study has since expanded into phase II of the investigation of the subjects involved, and has broadened in scope. This manuscript presents the current state of that investigation and goes on to examine the implications for the future of space flight under dramatically new conditions.

The first chapter will give an overview of the space elevator concept, its history and the developments of science that now put the subject in a new context. I have tried to write it such that the reader is not required to have a degree in physics to understand it. Hopefully, I will have succeeded in including the definitions and explanation of technical terms, that the reader will need, within the material. References, and there are many, appear at the back of the book. These refer to the list of previous works that much of this study is based on.

The chapters following the first will address individual technical components of a space elevator and the challenges that come along with building and operating such a system. These chapters expand on the concept stated in Chapter 1 with additional background and explanation for the interested public and those that may wish to follow up on any of this work. Unfortunately, all of the various aspects of the space elevator are interwoven. Each component is affected by the design of the others and each new challenge to the survival or operation of the space elevator has repercussions throughout the entire system. The fallout is that the chapters will reference each other extensively including chapters later in the manuscript. I apologize for this and hope it won't deter any interested readers.

Later chapters explore the economics of the space elevator, its effects on the development of space travel and the wider implications for the future of our technological society. For those of you who enjoy science fiction, future technologies, and challenges I hope both the technical and practical discussions in this work will spark your interest. For those of you who are pragmatic and down-to-earth, please, examine the details of this work, and see if you are not convinced that there is an interesting development on our horizon.

If all goes well, what we present here will turn into engineering reality and it will mark a major mile stone in human development. Future generations will look back on this as they now do to the electric light, powered flight, nuclear energy and the computer chip. This time, the concept that "couldn't be done" will be the stepping stone for us and our children to create a truly space-faring human society.

Acknowledgements

And now as I begin the technical assault on the space elevator, I would like to acknowledge the support that has made this study possible. As I mentioned already, the funding for this work came from the NIAC and I thank them for stepping out on the limb. This is one of the more futuristic concepts that NIAC (or any part of NASA) supports and I hope that this work will be a worthy return on their investment.

In addition I would like to thank many of my colleagues and friends for their helpful discussions especially: Eric Westling, Michael Laine, Bob Cassanova (NIAC), Foster-Miller Inc., Carbon Nanotechnologies Inc., Rodney Andrews (Univ. of Kentucky), Chuck Rudiger (Lockheed Martin), T.Y. Lin International, David Raitt (ESA), Edward Wullschleger, Tecstar, L'Garde, Bob Weller, Kelan Huang, Truell Hyde, Russell Potter, Paul Keaton, Nanoledge, David Smitherman (MSFC), Arthur C. Clarke, Ron Forward (Tethers Unlimited), Bryan Laubscher (LANL), Leonard David (Space.com), TechTV, Joe Wisnovsky (Princeton University Press), Bill Davis, Bennett Link (Montana Univ.) and Carla Riedel (Montana State), Hal Bennett (Compower), Hui-Ming Cheng, Bob Fugate (Starfire Optical Observatory), Mike Edwards (Motorola), Richard Epstein (LANL), Brad Cooke (LANL), Bob Roussell-Dupre (LANL), Carl Sloan and Ted Stern (Composite Optics), Larry Mattson (TRW), Paul Emberly (Kvaerner), Eureka Scientific, Ronald Doctors, David Vaniman (LANL), Jim Distel (LANL), Mervyn Kellum Jr. (LANL), Katherine Gluvna, and of course my wife Vlada.

Chapter 1
A Space Elevator?

We are now into the fifth decade of the "Space Age" and most people, of the western cultures at least, are aware that we put satellites into orbits about the Earth. The kids that were in school when Sputnik went up, watched, through their teen years, the fiery launches of the Mercury and Gemini programs and as young adults the awesome, visceral power, of the Saturn V that took the first men to the Moon. Their children, in turn, have watched the Space Shuttle and its no less thunderous ascension into the heavens. Two generations now, and working on the third, are intuitively aware of modern space flight and, without needing any technical training, are also aware of the tremendous power and the rather large expense needed to put things into orbit.

All this common wisdom about something they have had no personal contact with; maybe TV is good for something after all. On the other hand, this same group of people has had personal experience with all the other facets of modern transportation, cars, trains and airplanes. They have some idea, at least to the first order of magnitude, of what it costs to travel; $1020 from New York to Honolulu might seem reasonable whereas $ 10,200 would not. But they probably wouldn't think of that $1020 trip in quite the same way the space engineers think of it; for an average 75 kilogram person, as $13.60 per kilogram (kg). That is the way they think of space travel because the numbers get so large, from $10,000 to $20,000 per kg to put things into orbit on the Space Shuttle, depending on who is doing the accounting.

But what if there was a way to put people and things into orbit for something close to the cost of airfare? Would that change the world? You bet it would! As we shall see, when our tale unfolds, that kind of cost change would affect the world in

Fig. 1.1: Scale drawing of the Earth with a space elevator. The cable itself would be invisible on this scale.

ways we can hardly imagine. It would be at least as great a change as the computer revolution we are now living through and probably as great a difference as the granddaddy of all shifts in human productivity and wealth, the industrial revolution itself. It would be, in a phrase, "Out of this World!"

What's in an orbit anyway?

Some of the common terms we will be dealing with are as follows:

LEO or Low Earth Orbit starts at about 185 kilometers altitude and extends to 400km. This is the normal range of the Space Shuttle. It takes a speed of 7.6 kilometers per second (km/s), (27,360km per hour) to stay in that orbit. This is also called the delta V (ΔV) or change in speed needed to get to orbit. This is fast, very fast, in Mach numbers, 1 equaling the speed of sound it is Mach 23, or 29 times faster than a 747. To get things into higher orbit we have to use additional stages. Either the shuttle takes up a satellite that has its own rocket to boost it higher after being dropped off, or one of the disposable rockets like an Atlas or Delta acts as first stage and the second stage plus satellite go on from there.

Most satellites that we expect to stay in orbit for a few years have to be placed above 500km because there are still faint wisps of atmosphere at the lower altitudes. Just enough molecules to run into, to steadily slow the satellite down, lowering its orbit until it spirals into the heavier atmosphere and becomes just another shooting star some night. The range for these "high" LEO orbits is from 500km to 1700km depending on the mission. As we shall see, this "traffic" will become troublesome. Altogether, there are about 8,000 manmade satellites, new, used and parts-there-of, larger than 10 centimeters (cm), in orbit, and another 100,000 objects down to one centimeter in size. And every single one of them is going to be a problem that we will have to deal with shortly.

GEO or Geosynchronous Earth Orbit, as the name implies is the orbit altitude where a satellite is synchronous with the Earth's rotation, it takes the same 24 hours to go around so it stays over the same spot on Earth, it's "in sync". That altitude is 35,785km. The worth of GEO orbits is that you get the same view of Earth all the time, the Earth appears stationary from the satellite's point of view. You see this every time you see a weather broadcast that shows a weather map. It used to be that we got the actual satellite picture, in black and white, and we still do, with color now, when there are hurricanes to talk about. Most of the time, what you see on TV is a computer enhanced overlay of the satellite picture with far more detail.

Plan B

So, it is "obvious", vis-a-vis space, that the way to get there, is by "the rockets red glare". But is that the only way? Rockets weren't always so obvious. The Chinese had rockets, powered by "gun powder" (before there were guns) for a long time and never thought of using them even as weapons let alone going to the Moon. Not that they could have, but they didn't even think about it in stories and speculation; they just didn't have the concept of "Science Fiction". One of the earliest science fiction (SF) writers was Sir Isaac Newton. After figuring out the law of gravity, and hence what kept the Moon up and what an orbit was, he wrote about using a cannon (from a mountain top) and his third law, the one about equal and opposite reactions, to make a cannon ball an artificial satellite. As he was an expert experimentalist, it is a bit strange that he made no mention of how he expected to know if he were successful. Cannons remained a favorite speculation right up to Jules Verne using one in his story "From the Earth to the Moon" in 1865. But, by this time scientists could figure out that for the acceleration needed, somewhere in the neighborhood of 22,000 g's (gravities), a cannon would destroy itself and anything in it. There being no known energy source dense enough, (energy output, per unit mass) space flight was therefore considered impossible. But not for long.

In 1895 Konstantin Tsiolkovsky, a brilliant, self taught, Russian science speculator was the first to show that liquid fueled rockets would have the energy to do it. Not how to build the rocket, mind you, that would have to await Goddard in the 1920's, but that liquid fuels if burned efficiently had the energy to weight ratio to get to space. So we are back to rockets, yes, but that is not the news. Good old Konstant' also thought of a lot of other spacey things, he was the first to mention space stations in geosynchronous orbits, and by 1903, in articles, talked about a tower, up to geo-sync orbit and well beyond. He was the first to identify the concept that the part of the tower beyond geo-sync orbit would have an outward "force" due to Earth's rotation that would support the portion of the tower below geo-sync altitude.

But nobody listened, not even to Goddard, until World War II, then everybody was in the rocket business, and it became very apparent just how much rocket power and expense it would take to get into space.

From Hans Moravec (more on him below) we learn of the thought experiments of his friend and colleague Dr. John McCarthy who, in the early 1950's, came up with Plan B. In memos he described a synchronous Earth "Skyhook" going up to a space station in geosynchronous orbit. The dimensions of the thing were staggering. The cable would need to be

144,000km long, way past the 35,785km of synchronous altitude in order
to balance off (by centrifugal force) that portion going down to Earth.
(The same thing Tsiolkovsky thought of but no one in the West knew
about.) It would certainly be the longest object ever built and it has to
hang there and at least support its own weight. This would require that the
cable be tapered, the thickest part at synchronous altitude, tapering both
down to Earth and out to the end.

The taper must be enough to support the weight of cable portion
yet below it. He showed that this "taper factor" would be heavily
dependent on the ratio of the materials tensile strength to its density. He
then proved it couldn't be done with the materials then available. So he
invented the non-synchronous, rotating skyhook.   Smaller, shorter,
lighter, this version only needed to be about 10,000km long and its center
of mass would be in a much closer orbit, about 5,000km altitude. As its
center of mass passed around the earth as a satellite normally would, it
rotated so that the tip of each arm would touch down at the earth's'
surface, just as the spokes of a wheel would "touch down" on the surface
it was rolling on. He then proved that couldn't be done either. Decided to
found the Stanford Artificial Intelligence Lab instead. Plan B was not
looking so good.

Graphite Whiskers (fibers) invented 1957. New ball game.

Up till now the best materials anyone had to work with was fine
grade drawn steel wire with a tensile strength (TS) in the 42,000kg per
square cm (kg/cm^2) range and a density (d) of 7.8 g/cm^3. The tensile
strength tells us how much weight the material can hold up in a one
(Earth) gravity field (1g) for a given thickness of material, in this case one
square centimeter. Given the density of the material, this also tells us how
much of itself it can support, its self-support length; just divide the tensile
strength by the density. Steel wire, then, has a self-support length of 53.8
kilometers.

This is far short of getting to 35,785km, which means that as you
go up from the earth the cable size has to be increased periodically. This
is the taper ratio that McCarthy figured out. The actual amount of taper
for any point along the cable can be computed with a calculus expression
but there is a simple way to visualize what is involved. As a first
approximation you would have to double the size of the cable at the
halfway point of the self-support length. For our steel wire, every 27km
in order to support the weight below it. What we find is that the actual
number of doublings would be 183 times. Two to the 183rd power, of any
width you start with, no matter how small, is just as good as infinity for
the width at the top.   The actual taper figure that McCarthy got was

$1x10^{50}$, an impossibly large number, (wider than the solar system). (This is interesting, as we shall see shortly. It indicates that he probably used the same sort of step-wise integration that we just did.) As McCarthy concluded, "forget it".

But only a few years after McCarthy closed the case of the "too massive skyhook" new evidence appeared and a new suspect. With the invention of graphite whiskers we now have a material with a tensile strength of $210,000kg/cm^2$ and density of only 2.0 g/cm^3, roughly 20 times better than steel wire.

Now the ideas came thick and fast under a variety of names, beanstalks, orbital towers, skyhooks, space elevators etc.

Y.N. Artsutanov, Russia 1960, was the first to work on the problem using graphite whiskers. A fellow countryman of Tsiolkovsky, and champion of his prescience, he revisited many of his ideas and added sound mathematical foundation. He proposed the idea of building the cable from a geo-sync satellite in both directions, Earthward and outward, in such a way as to keep everything in balance and in orbit.

John Isaacs, Allyn Vine, Hugh Bradner and George Bachus, "Science", Woods Hole 1966. Not being followers of old Russian Science Fiction or their technical papers either they reinvented the idea all over again.

Jerome Pearson, 1975 "The Orbital Tower", article, "Acta Astronautica" déjà vu all over again; and again, in a 1976 article, added technical details, discovered cable oscillations or the whip-lash problem; and again, in 1977-78 articles, decided to do something different, Moon synchronous skyhooks, using L1 and L2 points. Overall he did add considerable math and engineering details to the discussion and we are still using them today.

Hans Moravec, 1977 technical paper, popular version 1978 article, "Cable Cars in the Sky" in Jerry Pournelles' "The Endless Frontier". Reviewed the literature, mentioned many of the above, credits John McCarthy (above). Went on to show that with the new materials, graphite whiskers, a synchronous (stationary) skyhook, although quite massive, could work. Showed McCarthy's non-synchronous rolling skyhook would also work with new materials and expanded that idea to several other innovative uses.

Arthur C. Clarke, "The Fountains of Paradise". 1978 fiction. Clarke had been reading everybody, especially Isaacs 1966 and Artsutanov 1960, above. His concept involved a two stage plan. First he invents super strong monofilament carbon fibers (!!) and built a cluster of cables, then he built a solid tower around the skeleton cables.(??) And he ran the, now Orbital Tower, just a little way past GEO to a massive counterweight. This design would limit using the system for throwing payloads to other planets (another concept we will address) but it still made a great book.

Charles Sheffield, "The Web Between the Worlds" 1979 fiction. [Now out in reprint, 2001] Published just months after Clark, they only learned about each other's books at the last moment. Clark wrote a preface for Sheffield's book explaining that neither copied the other, just an "idea whose time had come". That friendly commentary was needed because in addition to the "beanstalk" idea, both stories followed similar themes?!

Arthur C. Clarke, (again) 1979. "The Space Elevator:" A science article elaborating on the ideas he used in "The Fountains of Paradise".

With graphite whiskers, a tensile strength of 210,000kg/cm^2 and density of only 2.0 g/cm^3 we have a self-support length of 1050km, almost 20 times better than steel wire, but still nowhere close to 35,785km. We also learned that by itself that is not the key number to be concerned with. The self-support number is only meaningful if a uniform one gravity field could extend upwards for that distance, which it does not. Gravity gets less intense as we move further from the center of the Earth; it does so quite rapidly as the square of the distance. So, a wire could actually support a length of itself greater than the self-support number. For just planning purposes we don't have to figure out what that extra distance is, we can work it backwards. We can collapse the 35,785km down to its equivalent height as if it were all in 1g. That length is 4940km. That's only one seventh of synchronous altitude, gravity really does drop off quite a bit with altitude. (This "lightness" does not show up much in the first 12km us ordinary Earthlings are likely to encounter, but above that it starts to be noticeable. This is not the same thing as the "zero gravity" of being in orbit, this is just less gravity by being further from the center of mass.)

Now we can work more easily with the self-support length. Graphite, at 1050km self-support length is still short of the 4940km that now represents the distance to geo-sync altitude, so we still need a tapered cable. But it is already obvious that we are in a much better bargaining position with the laws of physics. As it turns out the taper ratio we need is only 100. A graphite cable with an area of 100 square centimeters at the top would have an area of 1 cm^2 at the bottom and be able to lift a load of about 100 tons to orbit. It would also mass over 700,000 tons and pose a certain hernia problem in getting that much mass up to geo-sync orbit.

Several of the writers in this period proposed similar solutions for this problem; go get an asteroid. The idea is to use the asteroid to solve two cable-building problems, sort of two birds with one asteroid. One is to get the large amount of material (carbon) needed to build it and secondly to shorten the length of cable you have to build by putting a mass on the end of the cable as a counterweight, see fig 1.2. (The cable still has to extend well past geo-sync orbit but not the full length 144,000km.) This makes for a great science fiction story, but creates three additional problems.

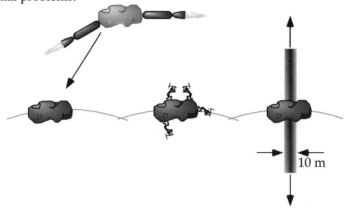

Fig. 1.2: Illustration of how a space elevator is constructed in several science fiction books. First, an asteroid is captured and placed in Earth orbit. Second, the asteroid is mined for its carbon. Third, a large cable is extruded both upward and downward until complete.

One is the difficulty of setting up manufacturing facilities, and large ones at that, in high Earth orbit. The cost: years and many tons. The second is getting the asteroid to cooperate. Moving one into Earth orbit is not easy. The cost: years and many tons (billions, see section on Mars cable). The third problem involves type casting, wrong hero for the role. Asteroids (including for the moment comets) come in two types. There are the rocky types, lots of metal content, solid, the type you need for a counterweight. Then there are the carbonaceous types, also called former comets. These are essentially slush balls, made up of carbon compounds and ice compounds, materials that you need to make the cable, but they would break up if used as a counterweight. Finding, then retrieving one that had a carbonaceous outside and a rocky inside would cost years and many tons. In short it would most likely cost more years and more tons than just doing it from Earth.

One of the most interesting essays in this period and worthy of mention because of what happened next was Dr. Charles Sheffield's, (again) in "Far Frontiers" 1986, spring issue. Good review of concepts, reviewed many of the above writers, added interesting details and broadened discussion to "dynamic" systems. Then, in what would prove prophetic, he bemoaned the lack of strong enough materials, so he invents hypothetical materials from physics principles with god-awful strength, shows how easy it would be to build a cable with them. Then he predicts technical advances would provide strong enough real materials by year 2000! (Good show, Dr. Sheffield! Nanotubes 1991). He has worked in background with many other scientists analyzing the physics of such systems.

Nanotubes invented, 1991, but the realization didn't sink in for a while.

Kim Stanley Robinson, 1993. Red Mars. Still used Graphite Whiskers and Clarke's Skyhook ideas in this development-of-Mars story but then it came to ignominious end; hey that is no way to treat a cable.

Robert Forward, 1995 "Beanstalk" in "Indistinguishable from Magic". Again, interesting details. The book is a collection of Forward's science essays, all of which have individual copyright dates in the 1980's except for Beanstalk. This is interesting. Was Beanstalk written in 1995 or earlier? He still deals with the subject from the point of view of Graphite Fibers rather than the new conditions presented by nanotubes (1991). He does, however, refer to the best material being fibers of "perfect diamond crystal", but there being no known way to make them. Almost there, but not quite. Going on to use the graphite whiskers available he proposes to make a pilot cable, 6000 tons, one square millimeter at the Earth end, with deployment done by dropping it down from orbit. Then more cable would be lifted from Earth on the pilot cable and attached, until its size is increased 100 times. Very prescient, now we are getting real close to a workable idea.

Here is how the competition stands at the moment.

	Density (G/cm^3)	Tensile Strength (kg/cm^2)	Self support (km)
Steel Wire	7.8	42,000	54
Graphite Whisker	2.0	210,000	1,050
Nanotubes	1.3	1,327,000	10,204

No Contest!

Nanotubes are clearly the strongest known material by a wide margin and at a theoretical strength of over 3 million kg per square centimeter, probably the strongest material that can be produced. There are no other known molecular bonds stronger than this arrangement of carbon that would be a stable solid, so something better is not likely to be invented. The tensile strength is somewhat dependent on manufacturing techniques, and since they are just now learning those techniques we are using the more conservative, and easier to achieve strength figure of 1,327,000kg/cm^2. That is plenty good enough. As a more practical example, a 3 millimeter (mm) wide string could support 40,000kg, 40 tons. (We will use "tons" to mean metric tons, 1000kg.) Even without skyhooks this will certainly revolutionize the elevator business.

But the real telling number is the self support length, 10,204km is way over the 4940km equivalent 1 gravity length. This implies that the nanotube (NT) string would not need any taper at all! A huge simplification, one string can run the entire length and still support itself. That mathematical point having been made, we will adopt the better engineering approach and explore taper factors between 1.5/1 and 2.3/1 in our considerations. The final design factor being a trade off between safety margin requirements for a given lift capacity and saving cable mass.

Rather than the out of reach 700,000 tons for a graphite whiskers cable we are now down to the more reasonable 22 tons for a barely minimum pilot string. We can do that! Plan B is looking better.

The challenge

Edwards: "I became involved in this subject one bright sunny day when I read a statement (author to remain nameless) that a space elevator couldn't be built for at least 300 years. Being somewhat aware of the literature on the subject I thought this an awful thing to say, especially without any supporting arguments, facts or information to back up the 300 years number. This was as good as a thrown gauntlet to a physicist. While still working at Los Alamos National Laboratory (LANL) I started looking to find out why it couldn't be built for 300 years, I'm still looking. I soon compiled a larger body of information on building a space elevator than anybody had before. I decided to put this into a paper (*Acta Astronautica*, Edwards 2000) and then decided 'what the heck why not send in a proposal to do more work on this?' I sent it to the NASA Institute of Advanced Concepts (NIAC) which often funds such studies. To my surprise NIAC accepted! They funded what is called a Phase I study, and I took a 6 month break from LANL to do official research on the Space Elevator. I was in heaven, pure research, no hassles for six months. I worked long hard hours because it was fun and in six months I produced the NIAC report, 'A Space Elevator' August 2000. NIAC was suitably impressed so I applied for a Phase II grant and NIAC funded that also. Now there would be the time and money to do a proper detailed study including a variety of investigations and actual experiments."

Here then is the better way, Plan B, thanks to modern materials and methods coming together in the 21st century.

If we assume for the moment that we can get all the carbon nanotubes we need to build a space elevator, we can build it in a way similar to how difficult bridges were built in the past. In building a bridge, the first thing that was done was a small string was thrown or shot across a canyon. Then a larger string is attached to this first small string and pulled across. This process is repeated, until many ropes and eventually structures are placed across the canyon. We have a serious canyon and the string is longer but the concept is the same. First, a satellite is sent up and it deploys a small "string" back down to Earth (see figure 1.3). To this string we attach a climber, at the Earth end, which ascends it to orbit. While the climber is ascending the "string" it is attaching a second string alongside the first to make it stronger. This process is repeated with progressively larger climbers until the "string" has been thickened to a cable, our space elevator. That's a pretty simple breakdown of what we are considering, although there are a few more details.

In considering the deployment of a space elevator we can break the problem into three largely independent stages: 1) Deploy a minimal cable, 2) Increase this minimal cable to a useful capability, and 3) Utilize the cable for accessing space.

The initial "string" we deploy from orbit is actually a very flat ribbon about 1 micron (1 millionth of a meter, 0.00004 inches) thick on average. It tapers from 13.5 cm (5.3 inches) at the Earth to 35.5 cm (14 inches) wide near the middle (GEO) and has a total length of 100,000km (62,000 miles). This ribbon, rolled up on a spool, and second stage rockets will be loaded on to a number of large rockets and placed in low-Earth orbit. Once assembled in orbit the second stage rockets will be used to take the ribbon up to geosynchronous orbit where it will be deployed. As the spacecraft deploys the ribbon downward the spacecraft will be moved outward to a higher orbit so that the center of this long assembly still crosses the geo-sync orbit to keep it stationary above a fixed point on Earth (a bit of physics we will explain later). Eventually the end of the ribbon will reach Earth where it will be retrieved and anchored to a movable platform. The spacecraft will deploy the remainder of the ribbon and drift outward to its final position as a counterweight on the end of the ribbon. This will complete deployment of a stable, small, initial ribbon under tension that can support about 1800kg (3970 pounds) before it breaks.

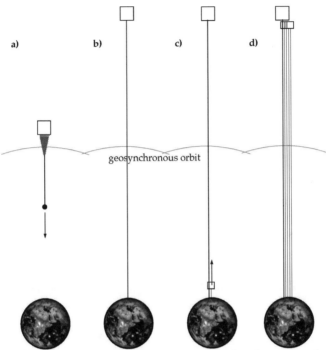

Fig 1.3: Illustration of the deployment scenario for the space elevator. A) A spacecraft is sent to geosynchronous orbit where it begins deploying a small ribbon. As the ribbon is deployed the spacecraft floats outward. B) When the end of the ribbon reaches Earth it is retrieved and anchored. C) Climbers are sent up the initial ribbon to strengthen it. D) A usable, high-capacity ribbon is completed.

The next stage is to increase this ribbon we just deployed to a useful size. During this stage climbers will ascend the ribbon and epoxy additional ribbons to the first one as they climb. At the far end of the ribbon the empty climbers themselves will become counterweights for the space elevator. One problem is how to get power to these climbers. Gasoline engines don't work well in space where there is no air and wouldn't have the required range, solar cells are too inefficient for their mass to be feasible, nuclear reactors are too heavy, an extension cord just simply wouldn't work, etc. The best option is to beam up the required energy. By using a large laser directed at "solar panels", photovoltaic cells, on the bottom of the climber, we can efficiently send up lots of power to the climbers. This power is easily converted to electricity for running electric motors to climb the ribbon. (See details in chapter 4.)

As each climber completes its ascent the ribbon would be 1.3% stronger. After 230 climbers (2 years), the ribbon would be capable of supporting a 20 ton climber with a 13 ton payload. This ribbon will have a cross sectional area, at the bottom, twenty two times the initial ribbon mentioned above. Payloads can be taken up the elevator to any Earth orbit or if released from the end of the ribbon, be thrown to Venus, Mars or the Asteroid belt. Initially, these payloads (large satellites, cargo, supplies, etc.) can be lifted about every three days. Additional ribbons of comparable capacity could be produced every 200 days using this first ribbon to string a new ribbon and "shipped" to other sites along the equator by dragging the lower end of the ribbon. In 4 years the capacity of any individual 20 ton ribbon could be built up to ten times that capacity, a 200 ton ribbon, and later even 1000 tons, a "1 million kilogram ribbon" with climbers roughly the size of an extended Boeing 747. Payloads as large as 700 tons could then be sent to Earth orbit, Venus, Mars or Asteroid belt every three days from one of these larger elevators.

The primary use of an initial 20 ton capacity ribbon may be to place spacecraft into low-Earth, through geosynchronous orbits. The uses for larger, 200 ton, up to 1000 ton capacity ribbons, as mentioned, would probably be for manned activities such as building and supplying a station at GEO or manned operations on the Moon or Mars. All of the lift costs for putting things in orbit from the space elevator would be a small fraction of what they currently are with rockets.

Now some of you have seen concepts for space elevators that entail grand designs, large futuristic transports, travel times of just hours to GEO, large city complexes at the ribbon anchor and complex systems with multiple tracks on single ribbons. The original science fiction concepts had these and it is a wonderful scenario to contemplate but they all lack a certain practicality, the real world of limited financial and physical resources. The design being proposed here probably sounds small, plain and boring when compared to these, but then science fiction concepts require science fiction budgets, we, on the other hand must start out more modestly. However, keep in mind that the first automobile was not a Porsche 911 and if man had refused to build any automobile unless it was a Porsche 911 horses would still be our primary mode of transportation today.

Murphy's say in all of this

We've all heard of Murphy's laws – what can go wrong, will go wrong. If we assume we can get the material to build the ribbon and that we can actually construct it as discussed above, are we home free? Not by a long shot. This is where Murphy has been working overtime. Getting the space elevator up is one thing, keeping it up there is something else.

The space environment is not a pleasant one; it's more like a burning and freezing, radioactive, corrosive, shooting gallery with no air. On top of that our own environment is not that pleasant at times with things like hurricanes and lightning. There is a whole set of environmental threats and practical problems that the space elevator will need to survive that were never included in the romantic portrayals depicted in the science fiction literature, including:

- Lightning
- Meteors
- Space debris
- Low-Earth-orbit objects
- Wind
- Atomic oxygen
- Induced Oscillations
- Electromagnetic fields
- Radiation
- Erosion of ribbon by sulfuric acid droplets in the upper atmosphere

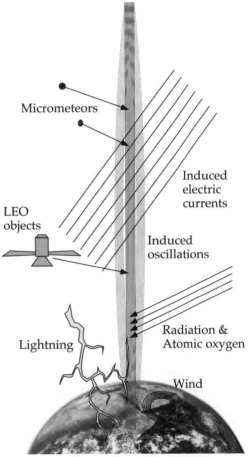

Most of these are capable of destroying our space elevator on short order if we aren't careful. The first lightning storm or strong wind would destroy the bottom end of the ribbon, meteors would shred it before we even got the initial ribbon deployed, atomic oxygen will eat it in a month whereas a low-Earth-orbit object would hit it every 250 days. Fortunately there are solutions to each of these problems.

What we will find and discuss in the following chapters is that each of the environmental problems will drive our design. We will also find that we are actually fairly lucky, there appear to be reasonable solutions to all of our problems. As you read through the initial chapters you may see design details that are driven by problems discussed later in the manuscript. These will be addressed more completely later in their own section. A few examples. The simplest cable design is round. Our cable design is a curved ribbon. The reason for not choosing the simplest design is that the round cable would be destroyed quickly by meteors whereas the curved ribbon is more robust.   Our anchor is also not a simple hook in the ground someplace in Kansas. Our anchor is located on a mobile, ocean-going platform in the equatorial Pacific, 2000km west of Ecuador. The reason for this is several fold. It turns out we can avoid the lightning and wind problems by locating our anchor at this specific point on Earth and by making the anchor mobile we can avoid collisions with satellites and debris in orbit.   Each of the problems we may encounter including stuck climbers and a possible severed ribbon will force us to a specific design.   In the end we find that we can solve all of the problems with a single and feasible design. We never found one killer problem that makes the design impossible.

Why would we want to build a space elevator?

Our society has changed dramatically in the last few decades from the first transistor to the Internet, DVD's and supercomputer laptops, from propeller airplanes to men on the moon, from hybrid plants to mapping human DNA. Often great advances in our society take a single, seemingly small step as a catalyst to start a cascade of progress. And just as often the cascade of progress is barely imagined when that first small step is taken. The space elevator could be a catalytic step in our history. We can speculate on many of the things that will result from construction of a space elevator but the reality of it will probably be much more.

At the moment we can at best speculate on the near-term returns of a space elevator. To make a good estimate of the returns we can expect we need to know where we are now, how the situation will change if we have an operational space elevator and what new possibilities this change will cultivate. First, where we are now:

1. Getting to space is very expensive: many millions for the launch of a small payload to low-Earth orbit, $500 million in launch costs to get a satellite to geosynchronous orbit and possibly hundreds of billions for a manned Mars exploration program.

2. Operating in space is risky. There are few situations where repair of broken hardware is possible and believe me launch shocks do break hardware. One hundred million dollars just for the insurance cost of a launch is not unusual. (The deductible – you don't even want to think about.)
3. Because of the limited, expensive access to space and the risk involved in space operations the satellites placed in space are expensive and complex. The satellites often cost more than the launch.
4. It is difficult to bring things back down from space. The only real exception to this is the space shuttle, and that only with objects it can reach, in LEO.
5. Neither the government nor the public accepts failure well in the space program. So rather than planning for ordinary engineering risks, whose costs can be rationally managed, (in engineering, failure is not a sin, only failure to learn from it) we spend a lot of extra money to avoid any failure at all, to cover "political risk".

That's the current situation. The next thing we need to know is how the situation will change if we have an operational space elevator. The space elevator will be able to:

1. Place heavy and fragile payloads in any Earth orbit (with a circularizing rocket) or send them to other planets.
2. Deliver payloads with minimal vibration.
3. Bring heavy and fragile payloads down from space.
4. Deliver payloads to space at a small fraction of current costs.
5. Send a payload into space or receive a payload from space every few days.
6. Be used to quickly produce additional ribbons or increase its own capacity.
7. Survive problems and failures and be repaired.

Having an operating space elevator would dramatically change our 'reality' picture of space operations as we described above. With this new set of parameters for space operations and the same economic reality we live in, we could reasonably expect the following in roughly chronological order:

1. Inexpensive delivery of satellites to space at 90% to possibly 99% reduction in the launch cost depending on the satellite, the orbit and the development of a multi-ribbon system. This would dramatically change the number of companies and countries with access to space and benefit from that access.

2. <u>Recovery and repair of malfunctioning spacecraft.</u> Telecommunications companies could retrieve and fix minor problems on large satellites instead of replacing the entire spacecraft. (Which now leaves an old dead satellite buzzing around up there in danger of hitting something.)

3. <u>Large-scale commercial manufacturing in microgravity space.</u> Higher quality materials and crystals could be manufactured allowing for improvements in everything from medicine to computer chips. These kinds of advances will never happen at current costs of space access.

4. <u>Inexpensive global satellite systems.</u> Global telephone and television systems would become much easier and less expensive to set-up and maintain. The reduced costs could make "local calls" to anyplace, but maybe Mars (at least initially).

5. <u>Sensitive global monitoring of the Earth and its environment</u> with much larger and more powerful satellites. Extensive private observing systems could be implemented to truly understand what we are doing to our environment.

6. <u>Large orbiting solar collectors for power generation and transmission to Earth.</u> This would truly shift the "balance of power" off the back of hydrocarbon consumption, allowing the development of limitless, clean power to the entire planet for private, commercial and vehicle use.

7. <u>Multiple, large and inexpensive spacecraft for solar system exploration.</u> Instead of very expensive small spacecraft taking a few photos we would have less expensive, larger spacecraft doing long-term planetary studies with videos, real time reporting and a suite of every valuable scientific instrument to fully understand our neighbors.

8. <u>Orbiting observatories and interferometers</u> many times more powerful than Hubble or any Earth-based radio telescope. Instruments many times the size of Hubble and gangs of interlocked telescopes multiplying their images a thousand times could search for and image planets around near-by stars. Spying out the future homes of Mankind.

9. <u>A manned space station at geosynchronous orbit</u> for research, satellite repair, commercial manufacturing operations and prep facility for deep space and solar system exploration probes. This would be a giant leap in man's occupation of space and it could come soon after construction of the first elevator. A large station (the size of a small town) could be placed in orbit and manned with a permanent crew (not only professional astronauts) doing valuable space work on satellites and research.

10. <u>Manned Mars exploration and colonization.</u> This is a large-scale occupation of Mars (hundreds of people) in the near future with a very

affordable budget. Only a space elevator system could support the frequency of traffic needed for such a project.

11. <u>Removal of man-made space debris in Earth orbit.</u> Our space debris is causing problems for satellites and this would allow us to clean it up on a realistic budget. Once you can inexpensively put satellites in orbit, and with a substantial fuel supply, you now have a "space patrol" and a variety of ways to find objects and slow them down enough that they will de-orbit.

12. <u>Spin-offs</u> would include inexpensive high-strength materials, high-power lasers, and high-purity and perfect structure materials. Better global weather monitoring, and prediction, and later weather changing capability.

13. <u>Military operations</u> would be dramatically altered with almost unlimited access to space. It's almost true now that "there is no place to hide" but the elevator would provide the inexpensive double and triple redundancy that would prevent an enemy ever "taking out" the surveillance system.

14. <u>Future mining of near Earth asteroids for valuable metals and water.</u> The third stage of space development, where space is the resource and Earth and the other planets are importers.

15. <u>Future vacation facilities in space.</u> This won't be tomorrow or even in the first ten years of operation of the space elevator but with an aggressive program and the change in the economy the elevator would bring, our children could make reservations for a week in orbit or on the Moon and be able to afford it.

These are some of the applications of the first space elevators and all but possibly the last two items would be feasible within the first fifteen years of operation including the manned exploration and colonization of Mars (see the section on destinations accessible with the space elevator). And again we believe these are feasible within the current economic environment when the commercial returns from the ribbon are factored in. Beyond fifteen years the best way to describe the impact of a space elevator is to say that we would have few limits in our solar system. For speculation on the possible long-term scenarios of space elevator operation we would suggest the ever vigilant and sometimes prescient community of science fictions writers, who, with their fingers poised at the keyboard this very moment, are ready to pour out fascinating speculation on this and many other futures that may await us. If you are in a hurry, and you like more science than fiction in your speculation, you can turn to the chapters at the end of this book, for more details of what the future might be like with the space elevator.

The Bottom Line

This is a feasibility report on the design and construction of a space elevator, it addresses all technical aspects of the problem from the deployment of the elevator to its survivability. This is not a definitive study or the final say on such a project but it is the most thorough attempt yet mounted to organize and deal with the details of the concept. What we have found is interesting. As we will discuss, building a space elevator will be challenging but not impossible, and so far, the more we study the concept the easier things get. For instance, our first cost estimates, allowing for all the unknowns at that time came in at $40 billion; a bit steep but for the advantages, well worth doing. Things change and in technology they can change fast. Progress has been made on the various technologies that have to come together to make this work and have lowered that estimate to under $10 billion, perhaps as low as $6 billion. This is not only less than many government projects it is within the reach of many commercial ventures! If the private sector decides to go to space, this is really the only economical way to do it. Again, it is hard to grasp the magnitude of the impact the space elevator would have on our society. We hope it will be clear from the discussion that follows that it would dramatically advance our society and the long-term return we would receive on such an investment would be staggering, it would literally change our world.

Chapter 2
Ribbon Design and Production

This is no mere rope, up close it might look like a length of package ribbon, although not quite so gay. Looking up it would appear rather small and innocuous, disappearing into the sky with no visible means of support. From a distance, if more than a couple of kilometers, you might not be able to see it at all. Nevertheless it does have substance. The design of the ribbon entails numerous considerations many of which we cover in more detail in other sections. In this section we will try to cover all the relevant components of the design and the influences that shape the design. We will begin with a background on the origin of carbon nanotubes, then look at the current status of nanotube progress and then address the specifics or our proposed design.

Any chemistry textbook published before 1985 would have told you, with sublime confidence, that there were two types of carbon. The two "allotropic" forms being diamond and graphite. Diamond, "the worlds hardest known substance" gets its properties by sharing its four electrons in a "tetrahedral" or box like pattern with four neighboring carbon atoms, forming a very strong 3-dimensional grid throughout the crystal. Graphite on the other hand is noticeably soft and brittle; we see this in pencils, it rubs off in an easily controlled way onto the paper. This also gives it the quality of being "smooth" and in very fine particles makes it a good lubricant. Graphite gets these properties from the carbon atoms arranging themselves in hexagon shapes, each atom attaching itself to three of its neighbors in a flat plane, instead of four in a box. This by itself would be a fairly strong bond; however, the weakness of graphite comes from these one molecule thick sheets attaching themselves only loosely to other layers above and below. These easily sheared layers make it graphite not diamond. Two forms of carbon; all is right with the world.

Bucky ball in the side pocket.

In 1985 Dr. Richard E. Smalley and Dr. Robert F. Curl of Rice University made a Nobel Prize winning discovery of carbon atoms that linked themselves together into a hollow ball molecule. A third kind of carbon! The atoms link together in the characteristic hexagonal pattern of chicken wire or more precisely, the alternating hexagonal then pentagon pattern of a soccer ball. More elegant comparisons were seen with the geodesic architectural style of Buckminster Fuller, and so, lucky for us, the name became "buckyballs" rather than chicken balls. In like fashion, the family of these compounds, of various sizes and shapes, is called "Fullerenes". Buckyballs are made by condensing, very hot (3000 – 4000K), carbon vapor. As the gaseous carbon atoms cool they start to revert to the solid state, this normally produces soot or "lampblack". But this time it was discovered that just as they first cool they form up into a geometric atomic arrangement that was not known before because it was too small to see. As the cooling continues the well-known clumpy forms of carbon, graphite, condense out, burying the cute little buckyballs so no one, until now, knew they were there.

The original buckyball, designated C60, was made up of 60 carbon atoms, the minimum number that can wraparound into a stable sphere. These chemically unique formations make hollow molecules a little over one nanometer (nm) wide. This is so small that we can't even call them microscopic, they are smaller than what optical microscopes can see, they are sub-microscopic! The newly discovered Fullerenes are now the third form of carbon bond and of the same strength as diamond bonding, thus they are often referred to as diamonoid material, or as one wag said, high class soot. (One is sorely tempted to extend the silk purse analogy to say that we are making diamonds out of soot, except for the small fact that, in this process, soot comes second.) With slight variations in the processing conditions different sizes and shapes result, so, labs all over the world started playing around with their buckyballs. Then in 1991 Dr. Sumio Iijima discovered tubular arrangements and buckytubes, now called nanotubes, were born.

Nanotubes, are 1 to 1.7 nanometers wide in a variety of lengths, open or closed ended, single, double or multi walled, with half of a buckyball sphere as end caps. They have several interesting properties. Just for starters, their tensile strength is 100 times stronger than steel at one fifth the weight. This is the highest rating of any known substance and, because of the way carbon atoms bond efficiently, is probably the highest it can ever get for a substance that can maintain solid form. Bundles of these nanotubes have been made up to 20 centimeters long (Zhu, 2002). This may not sound like a very long rope, but, consider, compared to the diameter, just a one centimeter length is the same ratio as a half-inch climbing rope that is over 400,000 feet long. And theory says that we can make them, not just centimeters long, but continuous, to any length!

For the next attribute we have to first correct a misperception. The pictures of nanotubes look like rolled up chicken wire, lots of open space, "airy" holes for things to pass through. A better picture would be to show these open hexagons as fuzzy areas, as if the chicken wire were made of fat knitting yarn who's edges spread out to fill up the openings. This would more accurately represent the chemical properties of these tubes, rolled up carbon graphite* sheets that are impervious to most other normal chemicals. Atoms can't easily fit between the carbon atoms, they bounce or slide off, most chemicals can't form a bond with the surface, they just loiter around. They use nitric acid to clean a new batch from the leftover impurities. This is not to say that nanotubes are invulnerable. They can be broken by cutting, or strong impact, and disrupted by harsh heat and some chemicals. But in the normal environment of air, moisture, and other outdoor travails encountered in everyday life they are far more durable than most of what we build things with today.

*[The sharp eyed reader will notice that nanotubes are rolled up sheets of graphite molecular arrangements, so why aren't nanotubes closer to graphite in properties (weak) than to the diamond like properties they have? This is because in being rolled up all the edges of the sheet are closed, (a closed crystal) there are no "loose ends". It is hard for a breakdown to get started.]

The importance of these three attributes, strength, length, and durability, is that we have the Ultimate Fiber! And now we have something really useful to do with it.

Status of Carbon Nanotube Development
The strength of nanotubes is impressive, wonderful for anything that you might want to make with these fibers but it is the strength combined with the low density of the material that is critically important when considering the design of a space elevator.

The strength of carbon nanotubes has been theorized to be as high as 300 giga-pascals (GPa or 300 billion Newtons per square meter). (If you prefer to think in terms of supporting weight, that would be equal to 3 million kilograms on a cable cross-section only one centimeter square. Or to put it even closer to home you could lift a large car with a nanotube "cable" the size of ordinary sewing thread.) That is an awesome sum compared to steel at only 4.2 GPa, Kevlar at 3.6 GPa, and graphite whiskers at 21 GPa. The density of the carbon nanotubes ($1.3g/cm^3$) is not only lower than steel ($7.9g/cm^3$) but also lower than its close cousins Kevlar ($1.44 g/cm^3$) and graphite whiskers ($2.0 g/cm^3$).

The critical importance of these properties is seen in that the taper ratio of the cable is extremely dependent on the strength to weight ratio of the material used. We demonstrated this relationship in chapter 1 with the self-support length, S/d = 54km for steel and 10,200km for nanotubes, or 188 times better than steel. (In our discussions the taper ratio refers to the cross-sectional area of the cable at geosynchronous altitude compared to the cross-sectional area of the cable at Earth surface. A taper in the cable is required to provide the necessary support strength to carry a load [Pearson, 1975]). For example, based on Pearson's work and operating at the breaking point, the taper ratio required for steel would be $1.7x10^{33}$, for Kevlar the ratio would be approximately $2.6x10^8$, and for carbon nanotubes the ratio is just 1.5. (Notice that McCarthy's value for steel, $1x10^{50}$ was a little off, but $1.7x10^{33}$ is no better, a one molecule wide string of steel, (and you can't get a smaller area than that) at the bottom would still be wider than the solar system at the top of that taper.)

Since the mass of the cable, to first order, is proportional to the taper ratio, carbon nanotubes dramatically improve the feasibility of producing a manageable size cable for a space elevator. In our discussions we have implemented a safety factor of two and are using a more readily obtainable 130 GPa tensile strength value. At this safety ratio, all points along the cable will have twice the strength needed to support the cable below it and the rated mass of the climber.

The momentum is building, carbon nanotube research has become a very active area with many hundreds of papers appearing in technical journals each year with dozens of research laboratories, private, government and academic popping up all over the world. This rapid progress in understanding the properties of carbon nanotubes and their production is very encouraging. We will briefly review part of that progress and what it implies, but keep in mind we can't possibly keep up when publishing a book, what we have written here will be out of date by the time you read this. In talking to researchers in this area, it is sounding sort of like a race to see who will make the ultimate material first. This is not a race to make space elevator cable, most of them have not even heard of our ideas as yet, it is the advantages to Earth bound engineering and technologies that are so attractive.

First, let's talk about the production of carbon nanotubes. Starting about three years ago researchers were starting to report growth of carbon nanotubes in aligned arrays up to a couple millimeters long [Ren; 1998, Choi; 2000] (figure 2.1). This was the first indication that there was a real understanding of how to grow carbon nanotubes in a controlled fashion. The laboratory production was not limited to a single method either. Nanotubes were being produced by electric arc and vapor deposition and variations on each. About the same time carbon nanotube ropes began appearing [Cheng, 1998] and have proceeded to become more impressive, up to 20 cm in length [Zhu, 2002] (figure 2.1). However, things have not stopped there, this is more likely just the beginning. Mitsui in Japan and Carbon Nanotechnologies Inc. in the United States have begun commercial production plans, tons of carbon nanotubes per week! (figure 2.1) A quick recap: carbon nanotubes discovered in 1991, macroscopic arrays and bundles by 1998, tons of carbon nanotubes per week production by 2003.   Twelve years from discovery to large-scale production.

A couple things we need to clarify for readers who want to examine the current status of the carbon nanotube development. There are two types of carbon nanotubes and they come in a few different forms. There are single-walled and multi-walled carbon nanotubes. This refers to whether there is one tube or if there are multiple tubes nested inside each other. Single-walled are currently better for structural use because it is difficult to attach to the inner tubes in a multi-walled nanotube. This may change in the future but a note to understand.

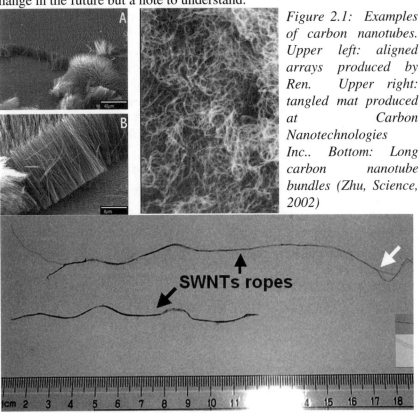

Figure 2.1: Examples of carbon nanotubes. Upper left: aligned arrays produced by Ren.   Upper right: tangled mat produced at   Carbon Nanotechnologies Inc.. Bottom: Long carbon   nanotube bundles (Zhu, Science, 2002)

Another key point is to understand the difference between an individual carbon nanotube and a carbon nanotube bundle. An individual carbon nanotube is a single molecule that could be nanometers to centimeters long. (Keep in mind that such a small object is hard to get hold of so that one can get good tests of property values.) A nanotube bundle is a string-like, macroscopic set of millions of individual nanotubes that are aligned and loosely held together by Van de Waals or other forces (think of it as surface tension, they stick together.).   The individual nanotubes in a bundle may run the entire length or may only extend a small fraction of the length. This is critical to understand. Several papers have reported the strength of a carbon nanotube bundle. This number is essentially meaningless. Pulling on two ends of a bundle is likely only to tell you how well the individual nanotubes stick together side to side and how they slide relative to each other but has nothing to do with the critical tensile strength of the individual nanotubes. To really understand the strength we have to know how many nanotubes extend the full length of the bundle and how long a bundle can be before you miss connecting, end to end, with a certain percentage of the fibers. (In other words, if you take up a shorter grip on a bundle you are testing you will probably get a higher strength number.)

One paper that did make the difficult measurement of the strength of individual carbon nanotubes was published by Yu [2000b]. In this and an earlier paper [Yu 2000a], Yu presents some of the only measured tensile strengths of nanotubes. Yu and colleagues appear to have done a thorough and well thought out experiment and got impressive results. Strengths of 11.2 to 64.3 GPa were measured for individual nanotubes compared to Yu's reference of theoretical strength of 300 GPa . (We used 130 GPa, Yakobson and Smalley, 1997, in all of our calculations.) The high measured strength in one of the first such experiments is encouraging.

In September 2001 we had a very cordial conversation with Mr. Ken Smith of CNI and got a look at their facilities. The ramp up to full production seems to be proceeding apace. With some details of the final design still not settled the cost of the production plant is expected to be in the range of $100 to 200 million, with capacity of 500 to 1000kg per day. With these facilities the cost should drop to the neighborhood of $1 per gram.

One of the difficulties of a new process, such as this, is efficiency, or rather the lack thereof. When we talked to Smith he estimated that the conversion rate of the carbon monoxide raw material to carbon atoms in a nanotube is about one in 10,000 atoms. This leaves a good deal of elbowroom for improvement in the process. Indeed, just one order of magnitude, improvement in throughput between now and the final tweaking of the production plant, (quite doable) and we are looking at a capacity of 5 to 10 tons a day by 2003. This would be more than adequate for our hypothetical order for 1000 tons a year, except for one thing. As Smith pointed out they are currently in the business of making nanotubes in the range of one half micron in length as a polymer to be used with, and to enhance the properties of, other polymers, such as adding electromagnetic properties to plastics. This is still a long way from making a 100,000km long ribbon.

These discussions gave us an excellent early look at the cost of our raw material. Let's say that another plant is built at the cost of $200 million (high end) for a production rate of 2000 tons a year (low end). Amortizing out the capital cost over 10 years gives an overhead cost per kg of tubes of $10. Multiply this by two or three times to cover operations and markup, as you see fit, and you have a cost to us of about $25 per kilogram, with fine prospects of much lower costs in the future. This puts the cost of materials for our first ribbon (before cost of manufacturing into a finished ribbon) at only $22 million! A very small sum considering our overall budget. This would seem to indicate that our main problem to overcome will be the manufacturing cost of the ribbon not the ribbon material. The progress at CNI also illustrates a very important change in concepts taking place. Right now nanotubes are a strange, wonderful, marvelous, almost science fiction like, exotic product. We don't want to have to build space elevators with exotic products, we want to build them with, soon to be, easily available, ordinary marvelous products!

Carbon Nanotubes in Composite

Having carbon nanotubes with a high tensile strength is only the first step, next is to build the ribbon with them. At this time our thinking is that the only realistic method is to incorporate the nanotubes into a composite. Having 100,000km long nanotubes woven or tied into a ribbon is not really viable for a number of reasons as we will see. The idea of making a nanotube composite is identical to making any composite such as fiberglass but nanotubes are a little different, their characteristics give them great attributes and create a few problems.

The primary difficulties in getting the carbon nanotubes into a composite matrix are dispersing the nanotubes and then getting a good interfacial adhesion between the nanotubes and the surrounding composite. First, nanotubes tend to stick together or clump. Various chemical and mechanical methods have been devised to separate the nanotubes. Second, the fact that nanotubes are essentially extremely long perfect crystals gives them their incredible strength but also makes them slippery guys to grab. By a method called functionalization, molecular "hooks" are incorporated into the nanotubes and these hooks can then be connected directly or through a matrix to transfer loads between nanotubes. Both the dispersion and the functionalization are being done at a number of institutions with initial success.

One recent development is the production of meter-length carbon nanotube composite fibers (10-50 micron diameter) and millimeter-wide ribbons (figure 2.2) by a French company (Nanoledge). The current stumbling block with these is their tensile strength. They currently have

0.5 GPa tensile strength. To construct the space elevator we will need at least 100 GPa and preferably 150 GPa or higher. Nanoledge is planning to have fibers of 5 GPa tensile strength within the coming year and start producing kilometer lengths.

Even more interesting is the file received the other day from Rodney Andrews at University of Kentucky. They have produced a carbon nanotube composite fiber 5km long that has 1% carbon nanotube doping. The introduction of the carbon nanotubes increased the fiber strength from 0.7GPa to 1.14GPa, which means they are getting good connections between the carbon nanotubes and the matrix. They still aren't up to what we need but they have made the first dramatic step. The next step is to increase the carbon nanotube component to 80% of the composite, which is possible in this scenario and improve the strength fractionally more.

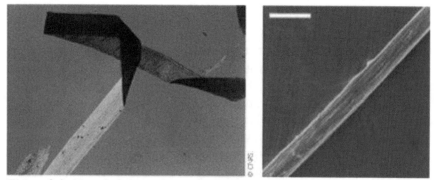

Figure 2.2: Nanoledge products. At left is a carbon nanotube composite ribbon illuminated with polarized light to show the alignment. At right is a carbon nanotube composite fiber.

As we are writing this more discoveries are occurring. This is the question, what will happen in the future and when will a high-performance carbon nanotube composite be in production. With the rate of developments that we have seen over the last couple of years, the dramatic increase in researchers in this field and estimates for milestones by the researchers we are working with, we believe that the technology required for construction of a space elevator ribbon will be ready by 2005 but possibly sooner. Specifically, a carbon nanotube composite fiber or ribbon with a tensile strength of greater than 100 GPa should be made within the next few years. Proper funding of a focused effort in this area would insure this development.

Now some readers are probably still saying, "Okay, but a fiber is still a long way from a 100,000km long ribbon." Not really, it is actually extremely close.

The Basic Ribbon Design (or Scotch tape and bailing wire)

Just kidding about the bailing wire, but good old Scotch tape is a surprisingly viable idea...

The ribbon of the space elevator is the most critical component and the most complex. The mass of the ribbon must be minimized while maximizing the strength along the axial direction. The ribbon must also easily survive meteors and debris up to centimeters in diameter, atomic oxygen, and wind. In an initial examination it appears that some of the requirements are incompatible but they're not.

To minimize the mass while maximizing the strength we need to have nanotube composites with parallel nanotubes and minimize all other mass in the ribbon. This means the strength is axially, as few cross or diagonal components as possible. Meteors and debris will destroy anything they hit, this can not be avoided. The ribbon must survive this damage. As we discuss later (Chapter 10.2) there are many small, micron size, meteors and debris but few, 2 cm and larger, objects to threaten the ribbon. To survive meteors and debris centimeters in size the ribbon must have one dimension much larger than this so only a fraction of the ribbon will be destroyed in any impact.

To minimize the mass while having one dimension much larger than centimeters implies a ribbon-type design, large in one dimension, very thin in the other. (This also gives a large surface area of uniform traction for the climbers.) This design will allow meteors and debris to impact the ribbon but not critically sever it, only punch small holes through it. The ribbon must also be designed such that the damage is limited to affecting only local areas. A design that meets these demands is shown in figure 2.3.

Our specific ribbon design consists of many individual fibers that are loosely interconnected (figure 2.3). The design has thousands of small diameter fibers (10 micron diameter) with cross-connections, or straps, across the ribbon at intervals of 10 cm or more. The cross-connections that we have studied are tape sandwiches that hold the fibers up to tensions of about 1 GPa for a 10 micron diameter fiber. Above this tension the fiber slips through the cross-connection. The result of this is that as a fiber is severed it contracts, pulling through the cross-connectors until the tension drops below 1 GPa at each cross-connection. When this happens the tension is transferred from the severed fiber to the neighbors through many cross-connectors over a length of 10 to 100 meters. The excess tension on the neighboring fibers is also transferred to its neighbors as it begins to stretch.

If multiple fibers are severed at one location then the cross-connection begin to slip on the intact fibers transferring the tension directly to several neighboring fibers. One thing that should be mentioned here is that a fiber under tension has a lot of energy stored in it. When the fiber is severed it will snap back like a cut rubber band. In our case there will be enough energy in this snap back to possibly do damage. Investigating the severity of this problem and ways to mitigate it is one of the important studies yet required.

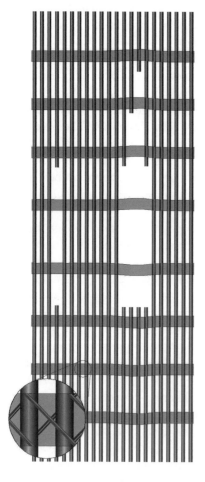

Figure 2.3: Proposed ribbon design. Fiber with carbon nanotube re-inforceed tape sandwich interconnects are shown (inset). Damaged areas with fibers which have slid are shown.

The individual fibers will need at least 100GPa breaking strength and will be used at 50GPa tension. The ribbon that we are proposing will have 3 mm^2 cross sectional area of 10 micron diameter fibers or roughly 30,000 fibers at the anchor. The proper adhesion strength for the cross-connectors will transfer about 1% of the individual fiber loads to the neighboring fibers. Standard adhesive tape has this performance. We took a standard, off the shelf, tape from 3M (Super Bond 396, recommended by 3M, Polyester taper, rubber-resin adhesive, 4.1 mil total thickness, 1.7 mil adhesive thickness, adhesion 190N/100 mm to steel) and sandwiched 7 micron diameter carbon fibers with tensile strength of 5GPa (Toray Carbon Fibers America, Inc., T700S). With as little as 2 millimeters of fiber in the tape sandwich we were able to hold the fibers to breaking. With thinner sandwiches the fibers pulled free. With a tape sandwich of one millimeter and 40% current commercial adhesion we would achieve the performance required in our ribbon. To demonstrate the design we have produced ribbons with nylon fibers and taped sandwiches. These ribbons were stretched to near breaking and then damaged intentionally. The severed fibers slid as expected redistributing their load (figure 2.4).

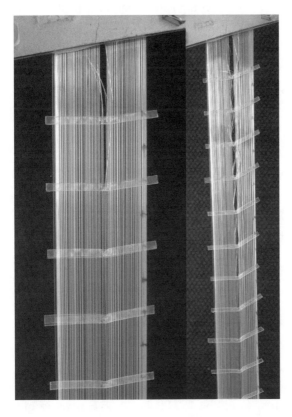

Figure 2.4: Segments of test ribbons constructed of nylon and commercial adhesive tape.

    Additional performance requirements for the tape include: sheer strength of 0.1N for a 1 millimeter tape width and 100 micron gap, resistance to radiation, resistance to vacuum, stiff holding adhesive, resistance to UV, and minimum mass. The first has been demonstrated in our tests as achievable. There are various adhesive, epoxies and backing materials that meet each of the rest of the performance requirements. However, additional development is required to find the optimal backing/adhesive combination that meets all of the performance characteristics. We have held discussions with 3M about this problem and are optimistic that the ideal tape sandwich exists currently in other applications.

Just as an example let's consider metalized kapton tape. Kapton has properties as shown in Table 2.1. Kapton tape is commercially made in various thicknesses including 7.5 microns. If we place 7.5 micron thick Kapton tape on both sides of the ribbon every 20 cm with a width of 1 mm we would have a total mass of metalized Kapton equal to roughly 10% of the total ribbon mass. Kapton appears to be a good backing for the space environment if metalized but an optimal and lighter mass substitute may be possible by using a carbon nanotube composite material. The adhesive itself also needs to be carefully considered. Adhesives and epoxies exist that do function in vacuum and radiation environments; a bit of a workout in the lab ought to come up with the right combination.

Table 2.1: Properties of Kapton

Physical Property	Typical Value at	
	23C (73F)	200C (392F)
Ultimate Tensile Strength, MPa (psi)	231 (33,500)	139 (20,000)
Yield Point at 3%, MPa (psi)	69 (10,000)	41 (6000)
Stress to Produce 5% Elongation, MPa (psi)	90 (13,000)	61 (9000)
Ultimate Elongation, %	72	83
Tensile Modulus, GPa (psi)	2.5 (370,000)	2.0 (290,000)
Impact Strength, N cm (ft lb)	78	(0.58)
Tear Strength—Propagating (Elmendorf), N (lbf)	0.07 (0.02)	
Tear Strength—Initial (Graves), N (lbf)	7.2 (1.6)	
Density, g/cc or g/mL	1.42	

Large-scale structure of the ribbon

The large-scale structure of the ribbon depends basically on the physics of a space elevator and the tensions that the ribbon must support [Pearson, 1975]. The overall shape is tapered on both ends and has its largest cross-sectional area at geosynchronous orbit

The length of the ribbon would be 150,000km if no counterweight were used [Pearson, 1975]. With a counterweight to shorten the upper end of the ribbon any length that reaches beyond geosynchronous orbit is theoretically possible. The shorter the ribbon the larger the counterweight mass required with it eventually reaching infinity when the ribbon only reaches geosynchronous. The interdependence of the total system mass, counterweight and ribbon length is shown in Edwards, 2000. The length of the ribbon should be determined by the counterweight available, ribbon size required, and the solar system destinations that are to be accessible from the ribbon (see Chapter 7: Destinations). In our proposed system we find that a ribbon 100,000km long is optimal from both a construction and destination stand point. This length will allow us to throw payloads to the asteroids, Mars, Venus, the Moon and almost to Jupiter. From the spreadsheet calculations of the ribbon taper we also find that the ratio of ribbon mass to counterweight mass, which is fixed by the physics, is 1.364. This ratio defines the masses of ribbons, spacecraft and climbers in our system and part of the reason this length was selected is that the 1.364 ratio is very reasonable for various components. The 1.364 ratio means that the ribbon that we will initially deploy will have a mass 1.364 times the dry mass of the spacecraft since the space craft becomes the counterweight. In the same respect the mass of the ribbon added to the initial ribbon would be 1.364 times the mass of the climber carrying it.

We have referred to Pearson's 1975 work before, but what you haven't seen, as yet, is his rather brilliant derivation of the ribbon shape in one mathematical expression. Brilliant, yes, but so complex that as yet, (we are working on it) no one has solved it (so we don't feel so bad). Here it is for those of you who like really messy integrals.

$$M_c = \frac{\rho}{\sigma} \frac{mM_pG}{r_o^2} e^{\left(1 + \frac{\rho M_pG}{\sigma r_o} + \frac{3\rho}{2\sigma} \sqrt[3]{(M_pGw_p)^2} - \frac{3r_o}{2} \sqrt[3]{\frac{w_p^2}{M_pG} + \frac{r_o^3 w_p^2}{2M_pG}}\right)} \int_{r_o}^{r_t} e^{-\frac{\rho}{\sigma}\left(\frac{M_pG}{r} + \frac{1}{2}w_p^2 r^2\right)} dr$$

On seeing this, one colleague remarked with some alarm, "It's only a cable, for G__ sakes, how can it have so many factors!?" M_c is the mass of the ribbon and that is all we are going to tell you, here; if you want to wallow in this mess you can find a write up with variables defined in the addendum – good luck. For the sake of the math, the only way this expression will be of any usefulness is to boil this down to a simplified form that can actually take plug-in values and give results for various factors. We're working on it.

Our "out", of course, is to revert to the primitive. The whole purpose of the calculus (Newton and Leibniz, in the 1670's) is to get rid of those egregiously tedious step-wise summations of minute increments of changing variables that was the only way to handle dynamic variables before. But, when the calculus itself is "egregiously tedious" then it is back to basics. Fortunately, in the years since Pearson's work the personal computer and spreadsheet were invented, just perfect for stepwise integration, and, with a basic understanding of physics, don't freeze on me here, we can produce an accurate model of the ribbon quite easily.

The information we will need for our model is quite straight forward (table 2.2), despite the look of Pearson's integral. (These numbers are for those of you who would like to check up on us but they aren't required to understand what is happening.)

Table 2.2 Some basic numbers

Cable tensile strength	1.3×10^{11} N/m²	T
Cable density	1300kg/ m³	D
Gravitational constant	6.67×10^{-11} m³/s² kg	G
Mass of our baseline climber	20,000kg	$M_{climber}$ (M_c)
Mass of the Earth	5.9788×10^{24} kg	M_{Earth} (M_e)
Earth's angular velocity – how fast the Earth spins	7.292×10^{-5} s⁻¹ rad.	ω_{Earth} (ω)
Radius of the Earth	6378000 m	R_{Earth} (r)
Our chosen safety factor	2	S

At the start, all we are doing is supporting a load with a certain length of ribbon. For some length, say 1000 meters, there is no problem to figure the amount of ribbon needed. But then the next kilometer of ribbon has to hold up the load, the first kilometer of ribbon and this kilometer of ribbon. Again no problem, except that now the forces have started to change. With altitude, gravity pulling down is not quite so strong and a new factor has entered the picture. The Earth rotates, imparting an upward acceleration that offsets the effect of gravity. (The actual effect of gravity, what we call "weight" in everyday life would actually be higher if the Earth were not turning.) To put these two components into a useful form we must express them as functions of their location relative to the center of the Earth. The equation describing the accelerations on our ribbon is expressed as $a = M_e G/r^2 - r\omega^2$, where r is the distance from the center of the Earth. The first term is the gravity term ($M_e G/r^2$), it gives the standard 9.8m/s² if on the surface, and grows smaller with altitude. The second term ($-r\omega^2$) is the outward acceleration due to the rotation of the Earth and is therefore subtracted; it grows larger with altitude.

Now for the first segment of ribbon at the very bottom of the elevator it will simply be a climber hanging on a ribbon, but to be consistent we must use the above form of the forces all the way up. To be safe we will make the ribbon twice as thick as needed. So the cross sectional area (A_{ribbon}) of the ribbon at the bottom will be $A_{ribbon} = M_{climber} * a * S/T$, where $M_{climber}$ is the mass of the climber, a, is our acceleration from above, S is our safety factor (2) and T is the tensile strength of the carbon nanotube ribbon ($1.3e9$ N/m^2). This equation is simply the downward force ($F=M*a$) on the ribbon divided by the strength of the ribbon material and we double it (S) for safety.

If we now go up the ribbon 1000 meters, for example, we have to now also add on the mass of the ribbon below us so we end up with a slightly thicker ribbon. Up another 1000 meters and a little more ribbon mass below us and a little thicker ribbon. You get the idea. We start with the climber mass hanging on the bottom, as the first row of the spreadsheet, calculate the downward force and then the cross sectional area of the ribbon. The next row (working down the spreadsheet) adds the downward force from the ribbon below us to the previous downward force and calculates the cross sectional area for that point on the ribbon. This gets repeated until you get to the end of the ribbon. At the same time, each time you calculate the new ribbon size, you can calculate the mass of the ribbon (step size * area * density) and sum this as you go, to get the total mass of the ribbon.

What this looks like in a spreadsheet you can see in the addendum where we show a text version to display the format. The first column is the radius from the center of the Earth starting at the surface of the Earth (6378km) and stepping up to 150,000km. Since the force changes are not very rapid, we used a step size interval of 100km for reasonable accuracy and manageable spreadsheet size. (We also used much smaller intervals, until we ran out of memory, to test for accuracy.) The next column, all you need is the radius (column 1) and a few constants from above. Column 3 is the cross sectional area, which uses column 2 and starts with a tension from the specified climber (load). Column 4 is the ribbon mass, which uses column 3 and some constants. From this you can see the mass of the ribbon, how the forces on the ribbon change from downward to upward, at GEO, and the optimal taper. You can finish it off by cutting off the ribbon at some point and seeing how much counterweight you will need (tension/acceleration). This spreadsheet would also allow you to play with things like the safety factor to see how the mass changes. This spreadsheet can also be used to model ribbons for Mars, the Moon or any other body, you just need to input the proper constants. (If you want to see the numbers go wild you can try making ribbons with the various materials before there were nanotubes.)

Variations on the basic tapered design can be implemented within limits. Some modifications to our simple uniformly thick and standard taper will help with several problems we expect to encounter. The first modification we suggest is to counter wind drag in the lower atmosphere, altitudes below about 9 kilometers (figure 2.5). For example, let's say that the normal size for this kind of ribbon with a straightforward taper was

roughly 10 cm by 1 micron for the pilot ribbon at the lower end.   Then we
can reduce this ratio from 100,000 (the 10 cm by 1 micron) down to 4000
(2 cm by 5 microns) and keep the cross sectional area and strength of the
ribbon the same.   This reduces the wind drag for the part of the ribbon in
the Earth's atmosphere by a factor of five (see Subsection 10.4 Wind).
Another problem that immediately comes to mind is meteor damage.   The
highest risk zone for meteors is between 500 and 1700km (figure 2.5).   In
this case we would do an inverse of what we did for the wind problem, we
would increase the width profile by a factor of 2.   Making the ribbon
wider makes each meteor incident affect much less of the ribbon's area
and therefore less of its strength; in this case by a factor of 5 (see
Subsection 10.2:Meteors) while only increasing the total mass deployed
by 0.65%.

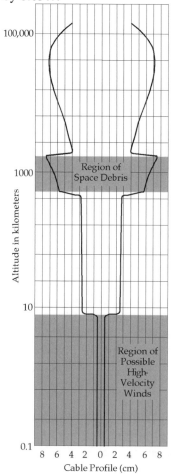

Figure 2.5: Large-scale profile of the proposed ribbon.

Figure   2.6:   The
Hoytether   design
(Robert Hoyt)

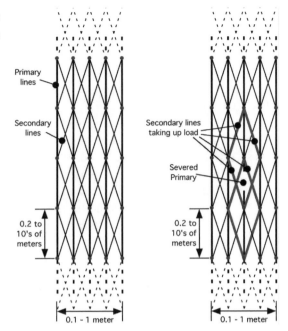

Alternative ribbon designs

Several alternative cable designs have been suggested. One is a design by Hoyt under a study for NASA's Institute for Advanced Concepts (figure 2.6). Hoyt's design of a space tether consists of both straight fibers under tension running the length of the cable and crossed diagonal fibers to take up and distribute the load in the case of meteor damage. Depending on the specific design, the Hoytether could be approximately 64% heavier (Hoyt's continuous high load tether) for the same load compared to our proposed design. In many tether applications this is not serious, in our system it would increase our system mass and launch mass by a factor of four. As we will see in the section on deploying the cable this could seriously hamper getting the cable launched with current technology. However, as published by Hoyt, this design may be more robust in terms of handling meteor damage and should be seriously considered at least for possible implementation in critical sections of the cable [Epstien, 2000].

Prior to deciding on a final cable design the problem must be examined in detail and segments of cables with various designs must be made and tested by shooting samples full of tiny holes. The critical parameters include the strength to mass ratio and the resistance to damage. They will be holed by meteoroids, the question of "resistance to damage" is not to be "bullet proof" but their ability to transfer load around the break and retain overall strength.

Ribbon Production

One of the major hurdles in the space elevator program will be production of the ribbons. The ribbons are unique in their design and have high performance requirements. Let's consider our design and starting points and then determine where the major production hurdles will be. To begin with we have:

- Nanotubes are beginning to be mass-produced.
- Nanotube composites are under development with initial composite fibers produced.
- Composite fibers are generally made through extrusion, which intrinsically aligns the carbon nanotubes.
- Fiber handling is a common practice in the textile industry.
- The tape cross-connects require additional engineering optimization but tape handling in general is also common in industry
- No defects in the ribbon are allowed.
- The length of the finished ribbon is to be 100,000km (see Chapter 7 Destinations).
- Hundreds of ribbons (ribbons to be added to the first to increase its strength) must be made in addition to the first pilot ribbon. Production time for each ribbon must be no more than one year and it must be possible to make up to 100 in parallel (see Subsection 10.2: Meteors and Chapter 12: Schedule).

The production of the ribbons would go something like this. Carbon nanotubes are mass-produced at one of several qualified facilities. The carbon nanotubes are dispersed and then implemented into a liquid composite. The composite is extruded into kilometer lengths of 10 micron diameter fibers and spooled. The fibers are run through fairly conventional textile processing which will align a set number of fibers in a parallel configuration. The fibers will be cross-connected with the optimized tape sandwiches. Spools of multi-kilometer long ribbons will be produced and then combined end-to-end to make the completed ribbons. This entire process will be completely automated. Each component of this process is achievable with current or near-term technology at the quantity levels we require.

Either during production or after the ribbon is complete several tests need to be done. These tests include:

- Integrity test: Insure that the ribbon has no holes or weak places. This may be done using optical scanning techniques, with the ribbon under tension.
- Spooling: Spooling and unspooling techniques must be examined prior to production of the final ribbons. It must be

insured that the ribbon can be spooled and launched without damage and unspooled without tangling.

- Tension tests: Spot tension tests may be useful but extensive tension testing (up to 75% of breaking) to insure quality will be required. For fine grain detail, only electron microscopes can see the individual nanotubes and since it would be prohibitive to scan the entire ribbon, statistical sampling would be used to evaluate nanotube behavior under load.

Chapter 3
Spacecraft and Climber Designs

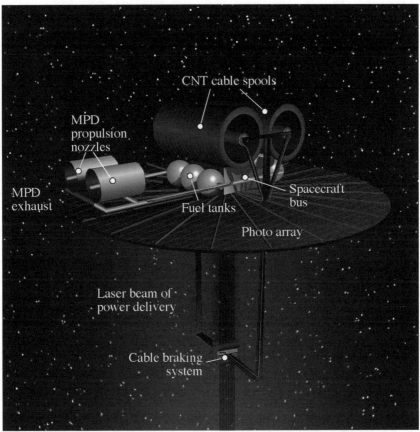

Fig. 3.1: Basic schematic of spacecraft to deploy the initial ribbon.

Initial Spacecraft

The spacecraft used to deploy the first ribbon will be very interesting and have some unique requirements. Let's compare it to regular spacecraft. There are lots of different kinds but typical are the telecommunications satellites that are launched into LEO and then need a second stage to boost them to their final orbit. Here they are set to rotating (given a certain angular momentum) so that they face toward the Earth all the time, unfold their solar panels to provide power and begin their task. Our spacecraft will do all this, just differently.

The first thing that is different is that the total spacecraft is so large that there is no way to launch it as a unit, we will have to send it up in pieces to LEO, assemble the parts and then move it out to its orbit at GEO. This will be the largest spacecraft ever launched from LEO (but considerably smaller that the International Space Station). The frame, to hold the other components, should be designed so that they "plug in" and are ready to go.

The final spacecraft leaving LEO will have quite a different engine system with long firing sequences at different times, and unlike conventional propulsion systems that expend themselves in minutes, ours must maintain operations and precise attitude control over several months. Even when it gets where it is going its job is not done. Its first stop is GEO, then it begins the deployment of the ribbon and at the same time continues to boost its own orbit on out to its eventual resting place at 100,000km from Earth where it takes up permanent residence as a counter weight.

This is actually the third configuration, or version, if you will, of the spacecraft. In the originally proposed scenario that we did for NASA in early 2000, the initial spacecraft was to be launched to geosynchronous orbit in pieces and four sections of ribbon would be assembled there for deployment of the pilot ribbon. When investigating a new concept such as this the first order of business is, can it be done at all, so anything that is feasible is Ok. Once things are "doable" then you can start worrying about "good, better, best". In the second configuration we found that a much smaller, simplified version with a single spool of ribbon, could be assembled in LEO and transported as a package to GEO. While this version was a significant improvement, the more we stared at it the more the disadvantages weighed on our minds. It would still take 7 or 8 launches to get everything up to LEO, putting the launch budget up into the several billions. Also, the second stage liquid rockets would be difficult, the fuel load is more than one launch could handle so fuel transfer would have to be done on orbit, in zero gravity and stored there for months. The mass was up to 186 tons and we were limited to a single 22 ton ribbon spool.

It was soon realized that the design criterion was the bottleneck. We had committed to using off the shelf technology that, at most, would only need minor adaptations; supposedly, because of the notorious high cost of developing anything new for space, this would save money and time. But with this "cheap" budget model approaching $4 billion just for the launch to LEO, one had to wonder if it wasn't worth looking at new technology, it could hardly be any more expensive. And there it was, a new system, Magneto-Plasma-Dynamic (MPD) propulsion, successfully tested in the lab, it looked like it could be developed into a system to meet our needs for a modest amount of money and only a little time.

The great advantage of this system is that the engines and fuel supply are light, only one tenth the mass of a liquid fuel system, Wow! The disadvantage is that this low thrust, MPD type engine needs an electrical power supply and it is very slow, months to get to GEO. As soon as we realized that the power system that we were already developing for the ribbon climbers was just the thing to power the MPD engines, we immediately decided that we weren't really in that much of a hurry. (Details below in the Propulsion section.)

Perhaps the biggest advantage is that the mass savings is so large that we can afford to send up two spools of ribbon, 20 tons each and get a pilot ribbon almost twice as large. And we can get all this on only four large launchers! This gives us a far more robust pilot ribbon, with less risk of failure to start with and a large saving of money and time in the subsequent build-up phase.

Thus the current configuration will consist of two ribbon spools, a mounting platform and propulsion/fuel system (we may go with one or a cluster of engines, to be determined by control of thrust factors). Other components will include a power system, communications, attitude control, navigation command, thermal control and structures to hold the ribbon spools and their deployment mechanisms.

From our deployment calculations we have determined some basic overall mass constraints (table 3.1), this is about half the mass of the previous plan, a tremendous saving on our launch budget. The masses for the subsystems can be estimated by comparison with existing spacecraft.

Table 3.1: Spacecraft Mass Breakdown

	Initial SC Mass (kg)	Requirements on subsystem
Payload (ribbon)	40,000	
Spools and structures	10,000	Support ribbon mass during launch
Power	500	Provide internal power for the spacecraft
Attitude control	300	Orient spacecraft relative to Nadir
Command	50	Control deployment, ribbon combining process, and ascent to far end of ribbon
Communications	50	Radios, video, transponders, antennas
Thermal	50	Maintain proper operating range
Spacecraft frame, panels, mechanical and other	2000	Mounting and mechanical system for deployment control, ribbon endmass
MPD engine system	2,000	Engines
Fuel tanks	2,000	Tanks with pumps
Additional counter weight	12,000	Can be used in systems above
MPD fuel	14,000	Fuel
TOTAL leaving LEO	83,000	Mass at start of MPD burn

Propulsion

As mentioned, for the orbit boosting trip from LEO to GEO, our latest design calls for dumping the tried and true liquid fuel chemical rockets in favor of the new Magneto-plasma-dynamic or MPD type electric engines. And we are going to save a huge amount of mass by leaving the power supply on the ground where access to power is unrestricted.

Of the many variations of electric drives in the works the MPD thrusters look most promising. The U.S. has now accumulated thousands of hours of lab tests so there is some track record to study. The Russians have tested a 200kW engine just 22 cm wide that generated 12.5 Newtons of thrust. It ran for 500 hours with no noticeable degradation. (We need something in the range of a year so the degradation would probably be noticeable, but at the end of the job that's ok.) We can take from this that the engine configuration we would use, an Isp of 4000 at 75% efficiency is well within reach.

These engines are deceptively simple systems, the engine chamber is a flared tube, the anode, with a large spike down the middle, the cathode. Fuel is injected into the system between the coaxial tube and spike where it is ionized. The large currents used in this system create electric and magnetic fields that propel the fuel ions out at very high speed, 40,000 m/s (an Isp of about 4000) in recent tests. This is actually a mass driver for very tiny masses. These drives use high current as their method, rather than the voltage based ion engines. These systems have the advantage over ion engines of not having to use a charged grid over the exhaust exit, which limits ion engine size and the amount of power that can be applied. Thus, MPD's have the potential of higher thrust, even higher efficiencies (Isp) and longer working life. We would like to have an engine life in the range of 10,000 hr.

The MPD is versatile, working over a range of 1000 to 11,000 Isp with a variety of fuels, ammonia (NH_3), hydrazine (N_2H_4), methane (CH_4), hydrogen (H), nitrogen (N), the noble gases (Ar, Ne, Xe), and alkali metals (Li, K, Na). It is also power hungry, with better Isp and efficiency the higher the power level, up to 80% to 85% efficiency with power in the megawatts should be obtainable; this would be in the thousands of ampere range. The lab people don't know yet how large they can scale this thing but they do drop hints of multi-kilo-ampere currents appropriate for multi-megawatt power plants. Fuel and power densities 100 times greater then the voltage ion engines translates to models that could use up to 100MW nuclear power plants at Isp's that may go above the 11,000 range. This puts them in the higher power class of devices that would be good for heavy cargo if time is not a factor. We need a small system compared to these monsters.

Their disadvantages are that they are very low thrust compared to chemical rockets, this makes them slow, and as just noted they need large power supplies that tend to be rather massive. They make up for the low thrust with very high fuel efficiency. In our particular example we plan to use an 800 kW engine (or its equivalent), with an Isp of 4000. The thrust is, $T = 2*(e)*P/V_e$, where e is the percent efficiency, P = power input and V_e = exhaust velocity, which also = Isp*9.8 which becomes $2*(.75)*800,000/39,200 = 31$ Newtons. Not much. It would take 138 days to go from LEO to GEO if we could use continuous burn, but the fuel efficiency is marvelous, it only takes 9.3 tons of fuel! Compare that to the 127 tons of fuel for the previous, chemical fuel engines. To put it another way, chemical fuel ships would move 0.46 tons of ship for each ton of fuel, same trip and payload. The MPD would move 7.9 tons of ship for each ton of fuel, a 17 to one advantage! (Theoretically, this could be expanded to a 1000 to one advantage!)

The other problem is the power source. Right now nuclear power plants are too massive for the power we need, although in the future that might not be true. NASA is working to develop a variety of nuclear fission reactors from 100 kW to 50MW for different mission requirements. From these studies, the power systems, including the supporting hardware and the propulsion engines, might get a weight ratio of about 10-kg per kW of plant size, e.g. the "specific mass" of the power plant is 10. A 1-megawatt power system (in the range we would need), a modest size, not too hard to do, would weigh about 10 tons, but that is in the indefinite future. We can do even better by not bringing along a power source at all. The plan is to beam the power to the spacecraft via a laser and receive the power with a Gallium Arsenide (GaAs) photovoltaic panel on board. (See power beaming section.)

The hardware that we will base the MPD engine design on is from the work of Edgar Coueiri of Princeton University and his colleagues in Russia. They have demonstrated a 200 kW MPD drive, and have achieved Isp's of 4000 using lithium.

Internal Power

Our deployment spacecraft will need continuous power for computers, electronics, communication and the mechanical controls for deploying the ribbons once on station. This internal power requirement is minimal, a few hundred watts, and can be accommodated through a conventional power system using batteries. The primary power system has the challenge of supplying the 800kW that will be required to power the MPD engines during transit from LEO to GEO and during the ribbon deployment. Any power requirements for the spacecraft (command, control, sensors, video, thermal) are small compared to the power consumed by the MPD and would be easily accommodated by the power beam. Once ribbon deployment has begun, the unwinding of the ribbon from the spool will generate several hundred kilowatts of additional electrical power that can supplement the power that is being beamed up to power the MPD engine.

Communications

The spacecraft will need an active, probably continuous, two-way communication package. All of the active equipment, computers, navigation, engines, fuel supply systems, condition sensors, etc. will have to be monitored and reported regularly to ground control. Even with the computers in charge of everything on board, these same systems will have to be able to receive and act on commands from the ground; this is going to be a long flight and it is doubtful that it would go completely automatically. In addition some type of active transponder for accurate tracking will be needed so that the power beam can target it and also a system of handshaking protocols to make sure that the beam stays on target. The data rates needed have not yet been determined but most likely we would use both wide-angle and directional antennae for flexibility. There should be nothing new here and readily available equipment should do just fine.

Attitude Control

Our initial spacecraft must present the same face toward Earth during its entire journey; this is called nadir oriented in the trade. This keeps the photovoltaic panels in view of the incoming power beam and also keeps the MPD engine, and its thrust, aligned in the direction of the orbital velocity. This is one of the most basic satellite orientations and requires Earth sensors, star sensors and a sun sensor as well as attitude control jets. Many satellites that observe Earth are nadir oriented, they just set up the proper rate of rotation for their orbit period and they stay pointed at Earth. Our system is just slightly different, each orbit is higher and thus longer in time requiring a new rotation rate every orbit.

The usual methods of attitude control are powered mass-gyroscopes, small attitude control jets and steerable main engine nozzles. Optimally we will use the steerable MPD engine to do the gross correction to the rotation rate while the propulsion burn is under way and a standard attitude control system for fine corrections. Once we get to GEO the task changes. To deploy the ribbon a small flyable "endmass" has been attached to the ribbon to help pull it off the spool and to control its direction. Once a few kilometers of ribbon are extended downward the system will be stabilized by gravity gradients and the spacecraft will be pulled into a rigid vertical orientation. Now the engines have to start moving the spacecraft out, away from GEO to offset the ribbon extending towards Earth. Since the descending ribbon is now holding the spacecraft in a fixed position relative to Earth the attitude control system only needs to keep the MPD engine pointed along the orbital velocity direction.

Spacecraft Structures

In discussions with Composite Optics (a company with experience in composite structures and systems for space) a spacecraft with 13:1 payload to structure mass is currently possible. Since our spacecraft, once in LEO, will not need to survive launch stresses, its structure can be lighter for the same payload mass as compared to current systems. Future developments and the possible use of carbon nanotubes (since they are required for the ribbon, we can assume that carbon nanotube composites will be developed prior to deployment of a space elevator) will push the payload mass to structure mass ratio even higher. Staying on the conservative side, our spacecraft design will use a total mass (not including the spent propulsion systems) to structure mass ratio of 13:1.

Counterweight

One would not normally add extra mass to a spaceship, and live to tell about it; but in this case we have to. It does nothing for the trip to GEO except cost us fuel, it is only there because once the ribbon is deployed the total mass of ribbon must be balanced off at the outer most end by a counterweight, at a specific ratio. This counter weight can be made up of anything that will remain stable such as all the parts of the spacecraft. In previous versions, using liquid rockets, there was more than enough mass, we might have had to drop something off to get the final ratio just right. The change to the MPD propulsion system has saved so much mass that we have to add some back on and take it with us. This will ultimately lower the overall cost of the spacecraft because we can avoid the usual costly fight to make everything as light as possible and it gives us the luxury of being able to include things that might otherwise be left out to conserve weight. If other additions are discovered during the actual design phase their mass can be subtracted from the counter weight allowance without effecting the total mass, fuel supply or length of the trip.

Ribbon Deployment

In our original space elevator system proposal [Edwards, 2000] both ends of the ribbon were deployed at the same time, the logical way to keep the center of mass at GEO. When the Earth end was tied down then the spacecraft moved outward to the end of the ribbon to act as a counterweight. However, deploying both ends of the ribbon from a spacecraft sitting at geosynchronous has some complexities. After further examination we believe that deploying only the lower end of the ribbon will work best. In this method, as the end of the ribbon is pulled downward the spacecraft can be maneuvered outward under power to maintain the overall center of mass at its geosynchronous orbit (see Chapter 5: Deployment). Eventually the end of the ribbon reaches Earth and the spacecraft continues to deploy more ribbon and floats outward to its end position (see figure 3.2). In this scenario the deployment of the ribbon, from a mechanical standpoint, is straightforward and has no high tension loading on the deployment mechanism on the spacecraft.

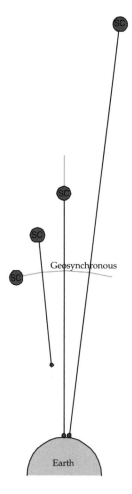

Fig. 3.2:Deployment of the initial ribbon. The ribbon is deployed downward (see discussion on imparting angular momentum to the ribbon) as the system is held in a geosynchronous orbit. SC = spacecraft

However, we are going to complicate the situation a little by having two spools of ribbon deploying simultaneously. The two ends will be joined together at a small "endmass" craft that has propulsion and directional controls to get the deployment properly under way. The mechanical system for deploying the ribbons will need to closely control the tension and speed of the deployment and insure that no tangling or twisting occurs. The computer will need to monitor the tension on each spool and keep them in sync, via the electric motors on each spool (used only briefly to drive and then for braking the rest of the time). In figure 3.1 the ribbons are shown as large-diameter core spools (3m x 2.75m O.D. x 1m I.D.), however, a longer spool (6m x 2m O.D. x 1m I.D.) should be investigated to reduce the chance of ribbon damage on launch.

With deployment speeds of 200km/hr and our proposed spool size we are talking about the spool rotating at less than 1000 RPM. This rotation rate is comparable to that of wheels on automobiles, the linear velocity is about four times as fast as commercial spooling machines. The spool itself should also be based on conventional designs where flat ribbons are spooled optimally onto a spindle with tapered ends. This allows the ribbon to be wound onto the spool with uniform tension across it at all times and an even, repetitive distribution on the spool.

The technique for beginning the deployment is interesting. There will be no forces initially pulling the ribbon away from the spools, so the little ribbon endmass craft will have to get things started in the right direction. This small craft has three purposes. The first is to impart a small amount of angular momentum to the ribbon as it is initially deployed. Once this initial angular moment is imparted and the ribbon is deployed to a kilometer or two of length, gravitational torque will keep the ribbon aligned, pointing toward the Earth. The second purpose of this craft is to transmit a beacon signal as the ribbon reaches the end of its deployment so the end of the ribbon can be found and retrieved on Earth. The third purpose is to have sensors to count the revolutions. With two ribbons hanging down side by side the tension in each ribbon would have to be perfectly equal for them not to rotate, twist, on the long trip to Earth. We can easily untwist them when they reach the ground if we know how many turns to take out.

Dissipating the Deployment Power

No power is required to deploy the ribbon (except for the first few kilometers) but a considerable amount of power will be generated by the spools as gravity drags out the ribbon. Depending on the deployment rate, the power that we need to dissipate could be roughly 300 kW for two weeks. This mechanical power can be converted to electrical with a system (DC electric motor and controller) having a mass of 105kg per

100kW and used to partially power the MPD engines during this phase of the operation.

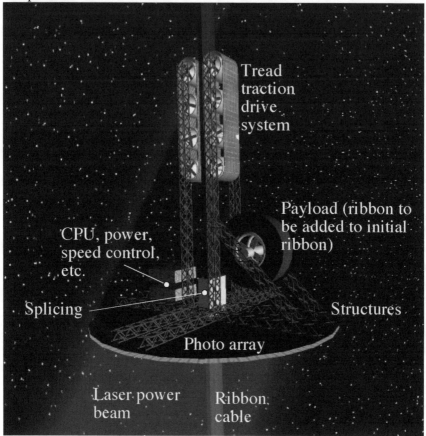

Fig. 3.3: Illustration of the various components of a climber.

Climber Design and Requirements

The climber will be designed similar to a spacecraft with some important differences. The mass, power, reliability and such are comparable to a spacecraft but the lift forces that the climber will be subjected to are minimal compared to that of a regular spacecraft during launch. However, unlike a spacecraft the climber will feel gravitational forces for most of its life. The climber will also have some unique mechanical requirements spacecraft generally do not encounter. The major components of the climber are the locomotion, ribbon deployment and power systems. There will be little 'thinking' to be done on the climber and minimal communications. The basic climber can be seen in figure 3.3.

From our deployment calculations we have determined a ratio of ribbon mass to climber mass for our proposed ribbon length and the lift capability of the initial ribbon. The ratio of ribbon mass to the rest of the climber mass is 1.364 and the total first climber mass is 900kg. This gives us 380.7kg for the climber structure and 519.3kg for the payload of ribbon. These mini-ribbons we will be deploying will each add additional lift capacity; 12.07kg for the first one. Each additional climber will carry slightly more ribbon than the one before as the main ribbon load capacity increases. The climbers themselves must also increase in size to accommodate the larger components. Some of the climber components do not need to expand linearly as the climber increases in size. This will allow us to increase the drive system faster than the climber and ribbon mass allowing us a faster climber. Manufacturing efficiencies will dictate at what intervals to increase climber dimensions.

Climber Ribbon Deployment

The primary job of the first 230 climbers will be to deploy ribbons as they climb and attach it to the existing ribbon. This will need to be done at high velocity (up to 200km/hr) with very high reliability and in no case damage the existing ribbon. The tension on the ribbon being deployed will need to be controlled carefully to insure there is no breakage and that the new segment is attached at a comparable tension to that on the existing ribbon. In our scenario we believe that these additional small ribbons should be added to the edge of the initial ribbon to widen it at least until the ribbon is roughly 30 centimeters wide. (The very first climber will be an exception. Its job will be to zip the two deployed ribbon halves together. One way to do this is by adding a string of new ribbon up the middle.) By widening the ribbon we reduce the likelihood of catastrophic meteor damage. Once the ribbon reaches a 30 centimeter width then additional ribbons should be used to thicken the ribbon. The final width of working ribbons may be a meter or more wide depending on the design needs of the traction system on the cargo version climbers, yet to be determined. The ribbon splicing system will attach additional ribbons to the first through tape sandwiches very similar to those discussed in the ribbon design chapter.

Motor Specifications

The locomotion system for our climber must be designed around the operational constraints of our entire program. The basic performance specs that we require include:
1. kilowatts to megawatts of mechanical power
2. high efficiency
3. high power to mass ratio

4. operation in air and vacuum
5. lifetimes of months continuous operation without fail
6. operate efficiently with torque and rpm ratios greater than 10:1
7. operate in a constant power / variable speed mode

The first six of these are pretty clear from a simple understanding of the space elevator. The last comes from the fact that our power beaming system will operate at a constant output. To best utilize the input power the climber's locomotion motor should run at a constant power. Since our load will be decreasing as the climber goes from a 1 g environment to zero g, a constant power implies that the speed of the motor will be constantly increasing during its ascent or a variable transmission is required.

Our motor study has come up with a motor design that fulfills all of the stated requirements. The motor would be based on permanent magnet brushless multipole technology to achieve a high efficiency with low mass. Cobalt-steel alloy and Neodynium-Iron-Boron magnets would be used along with a liquid cooling system and a two or three stage transmission. During most of the ascent these motors will run at greater than 96% efficiency and above 90% for most of the remainder. A 10kW motor of this design would have a mass of 14kg, require 5kg of control electronics and could be produced in quantity for under $9k. A 100kW motor of this design would have a mass of 105kg, require 20kg of control electronics and could be produced in quantity for under $50k. We would be restricted to using the small motors on the first climbers but the 100kW motors are obviously more efficient, both in mass allowance and cost per unit power, therefore we would want to use them as soon as possible in the larger climbers. As the climbers increase in size the drive system can be made a larger fraction of the climber mass (90% increase) allowing the early use of the 100kW motors. This will make the larger climbers 2.5 times as fast as the first one.

Track and roller system

The track and roller system, upon which literally everything hangs, must be designed to apply a smooth traction hold without damaging the ribbon. The frictional properties of carbon nanotubes are not known and the way they are grouped together in the final ribbon design may also change their surface texture. These factors will need to be examined before the track part of the locomotion system can be designed. In considering our ribbon design it is important that the track system grab the small structures of our ribbon uniformly. This would imply that the track in contact with the ribbon must be uniform and deformable on the micron scale. A tread or track system, which sandwiches the ribbon between it, is currently the baseline design.

This part of the system will also need to have a braking system in case power is interrupted and for the controlled "decent" after it passes geo-sync orbit. As we go past the zero gravity of geo-sync orbit we experience negative "gravity", the centrifugal acceleration of the rotating system makes it feel like we are going down hill, yet we are still going up the ribbon. The motors will no longer draw power, but rather will act as a braking system. The engineering of the climber must be such that it will be able to operate with all the forces on it reversed, going "head first" "down hill". The climber must also be equipped with a method to release the track from the ribbon externally in case there is a malfunction (see malfunctioning climbers below). Tests and experiments will be required to optimize these parts of the locomotion system.

In addition the tracking system must be designed to stay centered on the ribbon and not wander off to one side or the other. To do this the rollers which the track and ribbon passes through on the climber may be slightly narrower in the middle than on the ends. This would create centering forces that would keep the ribbon going through the middle of our roller system. An alternative active system with optical sensors and tilt control on the rollers could be used to maintain the alignment of the climber.

Power System

The climber half of the power transmission system consists of photovoltaic cells for receiving the incoming laser power and a power conditioning system. There are a couple options for the photovoltaic cells and the choice is dependent on the performance of the photovoltaic cells and available lasers (see laser beaming section). The most likely scenario is to have a 4m diameter array of photovoltaic cells located on the bottom of the climber. An alternative suggestion has been made (Hal Bennett) that the arrays could be located on the side of the climber and the power beaming stations could be fairly distant from the anchor (example: Mojave Desert). There are positive and negative aspects to this possibility and it should be examined further.

For optimal use of the motors it is best if our power system outputs voltages greater than 2500V. These voltages can cause arcing problems when passing through the upper atmosphere but this problem can be eliminated with proper engineering. In addition to receiving and utilizing the power beamed to the climber the power system must also be capable of dissipating excess energy that will be generated once the climber passes geosynchronous orbit. Depending on the velocity maintained, it will be necessary to dissipate kilowatts of power for up to several weeks. This can be achieved by radiating the power in one of several forms (light, RF, heat) or used in a cooling system.

Thermal issues

Another issue related to the climber that must be considered is thermal. If we consider a laser beaming system, we will have solar cells where possibly 10% of the incident energy will be converted to heat. For our initial climber this is 16 kW. In addition we will generate 1 - 4 kW of heat in the locomotion system. If the solar cells are isolated from the rest of the climber and exposed to space it is possible with our 4m diameter array the panels will come into equilibrium at 340 K, which is reasonable for photovoltaic operation. (This assumes a blackbody emissivity for the panels and a 293 K ambient environment on the under side of the arrays.) Since carbon composite structures are thermally conducting we may get additional radiative cooling simply by having good thermal connections to our structures. This will have to be investigated further. It may also be possible to design the photovoltaic arrays to further minimize the heat generated.

In addition there are thermal considerations in the motor design. Initially the motors will be operating at extremely low speeds which dramatically increases the heat production. This initial heat load affects the motor designs and can cause damage to the permanent magnets. To improve this situation we might consider a low-mass "tug" that takes the climber up the first few hundred meters until the climbers are easily pumped with the power beaming system and so the climbing motors would not need to start at zero velocity.

Structures

The support structures of the climber will require much less strength than that of a standard spacecraft since it will face no rocket like launch forces. The structure of the climber will be designed for a slowly varying load in the vertical direction with the primary structural loads existing between the ribbon spool tension and locomotion system. Design of the structures must consider thermal issues as well (see thermal discussion above). It is directly analogous to an elevator, just a high tech elevator that sometimes operates in a negative g environment.

Control system

The control system on the climbers will be required to monitor the speed of ascent, the tension in the ribbon, the splicing process and the climber location. During most of the climb the system will be in a slowly changing system with little complexity in the control required. Beyond geosynchronous, the climber will need to switch modes from a climbing mode to a braked descent. The final, and probably most important, responsibility of the climber controls is to stop the descent and lock the climber in place for use as a counterweight at the far end of the ribbon.

Communications

Communications, like control will be minimal. The only communications that will probably be present in the climbers are low baud rate diagnostics and emergency contacts. If the climber stalls, if there is a loss of power beaming or a problem on-board then the climber should send a prompt communication.

Mass Breakdown of the Climbers

Our construction climbers have two jobs, one is to lay ribbon the other is to stack themselves up at the end of the ribbon and become part of the counterweight system, a lonely job indeed. To do their second job properly they must maintain the right ratio for their empty weight, the counterweight ratio of 1.364 as mentioned above. For the first climber, massing 900kg this gives us 519.3kg of ribbon payload and 380.7kg climber mass to work with. This ratio must be maintained as each climber carries more ribbon on a slightly larger climber.

For doing what we need to do we have 380.7kg, without payload, for the first climber mass. The communications, and control system will require minimal mass (18kg each). The structures will have fewer constraints on them than on a spacecraft so we can use a value that is aggressive for a spacecraft (13:1 payload to structure ratio or 64kg). Thermal control is important in our situation but also fairly difficult to estimate the mass for it at this point. For the moment, we will assign 36kg to thermal control. If we assume we will have splicing interconnect material at 1% of the ribbon mass and a system to apply it including rollers we get a value for the splicing system of roughly 27kg. The actual control system for the power may be 27kg. A 100 kW climbing system (12 m^2 of photovoltaic arrays and five 20 kW motors) requires 148kg (21kg for the photovoltaic and 127kg for the motors). We will assign a value of 42kg to the track and rollers associated with the drive system. We have clumped wiring and associated hardware with the systems they are connecting. These numbers are not precise and will change when the system is designed, and proportionally as the climbers get large but they illustrate the basic mass breakdown of the system and what will be required. A summary of the mass breakdown is in table 3.2.

Table 3.2: Mass Breakdown for the First Climber

Component	Mass (kg)
Ribbon	520
Attitude Control	18
Command	18
Structure	64
Thermal Control	36
Ribbon Splicing	27
Power Control	27
Photovoltaic Arrays (12 m², 100 kW)	21
Motors (100 kW)	127
Track and Rollers	42
TOTAL	900

Expandable climber design

Since we will be sending up hundreds of climbers of varying sizes it is critically important the climber is designed to be expandable. The design must allow for larger ribbon spools, addition of motors, increasing the strength of the structures, additional photovoltaic panels, higher heat loads, and higher power flow (see figure 3.4). The control and communications systems are the only ones that will not expand as the climbers grow.

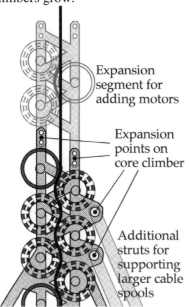

Expansion segment for adding motors

Expansion points on core climber

Additional struts for supporting larger cable spools

Fig. 3.4: Examples of expansion points that can be added to the climber in figure 3.3 that would allow for easy increase in climber size without a complete redesign.

Malfunctioning climbers

In the unfortunate case where a climber becomes stuck during its ascent there must be a method for removing the climber. One fact that complicates removal of stuck climbers is that the ribbon will probably not be able to support two climbers both at low altitudes. However, there are methods to get around this difficulty.

One option if a climber becomes stuck at a low altitude is to pull the ribbon down until the climber is retrieved and then allow the ribbon to float back out to its nominal position. In the current design of the ribbon the fraction of breaking strength would be pushed from 0.5 in nominal operation to 0.6 if 3000km of the ribbon were reeled in to retrieve the climber. The highest stress in this situation is at the far end of the ribbon; the rest of the ribbon would be at less than 0.6 of the breaking tension.

A second option presents itself above 2600km where the downward acceleration (gravity) on the climber is less than 0.5g. In this situation a second climber without payload could be sent up to release the malfunctioning climber and carry it beyond geosynchronous orbit where it could be released. If you really wanted to cut it fine, special rescue climbers could be made by stripping them of unneeded items, leaving a mass of about 342km, 38% of a full climber. This would drop the altitude requirement to 1640km staying within the 50% safety factor and all the way down to 630km if willing to go with a 40% safety factor for the short time it takes the rescue climber to gain a little altitude and have the gravity gradient lighten the load. All climbers would be equipped with a release that could be accessed by a climber coming up from below. Between 630km and 3000km either of these two options are viable.

A third option is to have the climbers equipped with release mechanisms that can be triggered by command from the ground, letting the climber just fall away. This has some hazards that must be investigated. One is to the ribbon being scraped by a falling climber, which might be remedied by a small emergency cold jet to thrust the climber away from the ribbon. The second hazard is to things on the ground, which might be remedied by good radar tracking. This should only be a problem for low altitude releases, above 3000km they should gain enough speed to burn up in the atmosphere.

Repair climbers

In addition to the standard climbers and a rescue climber a repair climber may also be warranted. This repair climber would be sent up with short sections of ribbon material that would be epoxied over weak sections. The climber would travel up the ribbon searching optically for weak sections in the ribbon and then place a ribbon patch on this section.

The difficulty is that to do this efficiently the patchwork would have to done at a speed of 100 to 200km/hr, unless we devised a sophisticated control system to stop, start and hunt at slower speeds. Repair climbers might also serve as rescue climbers if it were thought that the stuck climber may have damaged the ribbon. Repair climbers could also re-apply the coating (see chapter 10) to protect against atomic oxygen erosion. This could be a deposition process or possibly a metal-impregnated paint. The repair climbers would also be less massive than the ribbon carrying climbers and thus may be able to be sent up in between the standard climbers with minimal schedule impact.

Both repair and rescue climbers may need to be equipped with video cameras so we can see and evaluate situations on the ribbon. Today's video cameras can be very small; it's the transmitter that may cost some mass.

Climbing Stage

As we stated at the beginning of the climber discussion, after deploying the pilot ribbon it will need to be quickly strengthened and widened to make it more usable and more resilient to its environment. As the strength of the ribbon increases the mass of the climber and the ribbon it carries increase. As each climber reaches the 0.1g point (119 hours initially and 71 hours eventually)* the load on the ribbon has been reduced to a safe enough level that a second, slightly larger climber can be attached and started up the ribbon. Keeping to this schedule of new climber starts, a 20 ton capacity ribbon can be built in 2.5 years or a 200 ton capacity ribbon in 4 years. The mechanical power utilized by the climber primarily depends on the size of photovoltaic arrays and motors that can be carried. It may be possible to improve our current design and allow for more power per kilogram of climber (currently it is 100 kW for 900kg or 110W/kg). If we can improve our power to kilogram ratio by 40% (140kW for our initial climber or 155W/kg)** then we can strengthen our ribbon to 20 ton capacity in 1.5 years (saving 7 months). In addition to getting a large ribbon on-line faster the quicker schedule also reduces the risk of ribbon damage.

*[The gravity gradient is reduced to 10% of g at an altitude of 13,791km above the Earth, but the ribbon imparts a centripetal force on the climbers that reduces the effect of gravity, making the effective 10% g altitude 12,821km. Starting another climber at this point only loads the ribbon to 55% of capacity for a short time as all climbers continue to lose effective weight, from the ribbon's point of view, as they ascend. The 119 and 71 hour spacing of climbers is based on a working speed of 200km/h to cover the 12,821km distance, assuming that the slower speeds at the

start are later compensated for by higher speeds at lower gravity and reduced payload.]

**[Engineering milestones to look forward to in this regard would be points like the 54th climber, where all the ribbon and climber mass's have doubled giving us a better portion to devote to increasing the power to weight ratio.]

Chapter 4
Power Beaming

Getting enough power to a climber such that it can travel from Earth to geosynchronous orbit in a reasonable amount of time is one of the technological challenges of building and using a space elevator. When considering the situation we are discussing, we find the only realistic method for getting power to the climbers is to beam it up; that's the power not the climber.

Alternatives that have been suggested include running power up the ribbon, then using it to power the electric traction motors as you would for electric trains. Another is to use that electric current in the ribbon to inductively power some sort of magnetic levitation ("mag-lev") arrangement. The mag-lev has particular appeal because if you can suspend the car ever-so slightly away from the ribbon, (levitation, not touching it) you have the chance at least of engineering high rates of speed, far beyond what the system can handle with any contact drive. The idea of powering the ribbon, for either of these methods, probably comes from the fact that nanotubes are known to be very good conductors of electricity, perhaps the best, short of superconductors. But the scale is all wrong; even with very low resistance you can't send power over any "wire" for 100 thousand kilometers without horrendous line losses, thus acquiring an electric bill worse than buying fuel for the shuttle. Another point is, it would take two ribbons, (conductors) and you would have to guarantee that they would never swing together and touch for all that 100 thousand kilometer length.

Other suggestions have been for solar or nuclear power onboard or using the ribbon's movement in the environment's electromagnetic field. None of these methods are feasible due to efficiency or mass considerations. Solar panels, for the power density needed, would far out weigh the climber they were in charge of and they scale with the size of the climber so you would never catch up, the mass of panels plus climber would tend toward infinity. Nuclear power would be ideal, self contained, independent power giving the same advantages as cars and airplanes, the efficiency of onboard power and a much smaller external support system. But this is way in the future when we can do very compact reactors and very large climbers. And lastly, induced electrical power from the charged magnetic fields surrounding Earth is unrealistic. The magnetic and charged particle influences around the Earth are quite complex, made up of several fields and widely variable conditions, day side different from night side, Sun activity level, etc.

The short of it is that there isn't the flux density to do the job. Even though these fields are sometimes strong enough to knock out electrical equipment here on Earth, much of the time the fields are thousands of times weaker than the Earth's own magnetic field on the surface. A little intuitive engineering will put things in perspective. If we could suck power out of this type of magnetic flux we would be running cars on it and flying airplanes on it during strong aurora borealis. So, if you can't run cars on it down here you aren't going to up there.

There are two scenarios we have been considering for beaming power to the climbers, microwave and laser beaming. In this section we will go through the driving constraints of both systems. In examining power beaming we found both of these techniques have been considered in the literature as possible methods for transferring power across large distances. In the laser beaming case we even found a system under construction that meets our needs.

Laser Power Beaming

First, we will examine the laser beaming system. Our preliminary examination of this scenario suggested the power beaming station may need to be located at a high-altitude site (greater than 5km altitude) to get above a significant fraction of the Earth's atmosphere and thus be able to focus a beam tight enough to efficiently deliver power to a climber. High altitude operations are not impossible but do cause numerous difficulties and limitations. First among those limitations is that the ribbon is not likely to be located at the same site, since the requirements are quite different. Therefore redundant beaming stations would have to be installed at the ribbon site to power the climber to altitudes where it would be visible, over the horizon, from the high altitude site. Power beaming from sea level, where we would like to put the ribbon, is far more practical from an operational point of view, so we will examine this possibility in detail.

The primary difficulty in beaming laser power up to the climbers, specifically from sea level, is atmospheric distortion. Atmospheric distortion will broaden the beam and reduce the power delivery efficiency.

From *The Infrared and Electro-Optical Systems Handbook: Atmospheric Propagation of Radiation*, we find a discussion of exactly the problem we are investigating. The problem of sending a laser beam from Earth to space.

The long-term beam radius can be expressed as:

$$\left\langle r_L^2 \right\rangle = r_d^2 + \frac{4L^2}{\left(kr_o\right)^2}$$

where r_o is the transverse coherence length, L is the distance from the transmitter to the receiver, and r_d is a component of the diameter of the transmitter. This can be broken down to the radius of beam dispersion by two factors, broadening and drift or atmospheric wandering. (For an expansion of the math see Note 1 at end of this chapter.)

For a climber at 10,000km, with a 10m transmitter of 0.5 micron wavelength coherent light, we would have a broadening of 57m radius and drift of 9.3m radius. This is way too *broad* for our use. The beam would have about 20 times the radius (400 times the area) of the intended receiver, dissipating over 99% of its power elsewhere. With this spot size the wander is not significant but will be significant if the spot size is reduced. To beam up power from sea level we will either require adaptive optics, build unworkably large receivers, or we live with an efficiency of <0.25%. Can we use adaptive optics in this situation?

Adaptive optics (AO) has been used for a decade or more in large astronomical telescopes such as the 10m Keck telescopes in Hawaii. They have been very successful in eliminating the "twinkle" in star images, the wavering of light as it passes through the atmosphere. In our case we are doing the opposite, varying the beam to exactly cancel the distortion being caused. From the work of Robert Fugate and others we find that AO has experimentally demonstrated a spatial resolution of 25 cm at 1000km [Angel, 2000]. This is an order of magnitude better than our application requires at 1000km and this system can focus the laser into the precise spot size we need at 10,000km.

With this accuracy we can place the power we need onto the 3 meter diameter solar array we have designed into the smallest climber. By the time the beam expands to fill the photovoltaic array of our smallest climber (12,000km altitude) the power requirements of the climber are lower due to the reduced downward acceleration of gravity (~0.1g). In addition, at this altitude the next climber can start its ascent and the speed of the first climber is less critical (again reducing the power requirement).

Fugate and others have examined the problem of power beaming using lasers and find the same basic AO techniques work for power beaming that have worked for astronomical observing. They are currently planning a power beaming demonstration from Earth to a geosynchronous satellite [Lipinski, 1994]. The major problems that hinder the AO applications that Lipinski is working on would be the lack of a bright guide star and tracking moving satellites. We have neither of these problems in this application. The climbers will be at known, slowly-varying positions and a cooperative client that can be made to retroreflect part of the pump beam or emit a similar kind of tracking beacon. The only thing that has not been demonstrated is the complete beaming of a high-power laser. The primary problems that may be encountered in this next stage include thermal blooming of the atmosphere, and production of the high-power laser. In our application, thermal blooming, the overheating of a section of air the beam passes through, will not be a problem with a large beam size and the power we will be using.

A complete power beaming system with 200 kW of power is the aim of Compower, a private company [Bennett, 2000] (see figure 4.1). The laser power will come from a 200 kW free-electron laser (FEL) which the University of California - Berkeley will be supplying for a fixed price of $120M. The 0.84 micron (high inferred) output from this laser will be directed and focused with a 12m mirror based on the Hobby-Eberly telescope. The mirror is being modified to have more closely spaced actuators to accommodate the adaptive optics system. The system will deliver 200 kW (3 picosec pulses every 1 nanosec) into a 7m diameter spot at geosynchronous with up to 30% wallplug to laser power efficiency. The design of the laser is also readily expandable to 1MW. The current design only uses one fifth of the system (one of five wigglers) to produce the 200 kW. (Wigglers are sets of electromagnets that bend (wiggle) the electron beam and each time they do the beam produces light.) If all five wigglers are utilized 1 MW of laser power should be produced.

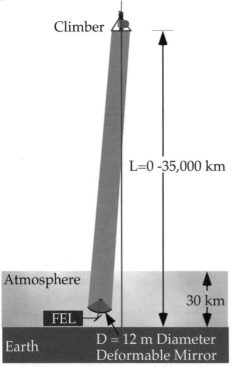

Fig. 4.1: Illustration of the laser beaming scenario.

This system has a five year construction schedule and its planned use is for delivering power to geostationary satellites.   To expand to higher powers either the system can be redesigned or multiple identical lasers can be used with their pulses interlaced in time. In our scenario we will need 2.4 MW of power delivered to our 20 ton climbers so three of these systems would need to be brought on-line and their pulses interlaced.   Recent progress is most encouraging, Hal Bennett, head of Compower reported that current tests show 350 kW per wiggler, for a potential full system output of 1.75MW per unit.   For the longer-term a $1x10^6$kg climber would require 120 MW of power delivered. We are only considering the first 20 ton capacity space elevator but the long-term aspects of the program must be kept in mind.   It is conceivable that subsystems such as the power beaming facility for the large climbers might become the schedule driver in construction of 200 ton, up to 1000 ton capacity space elevators.

The receiver system must also be considered when examining the power beaming system.   There are several photovoltaic cells that can be used as receivers and the choice depends on the laser being used, cell mass and the desired operational lifetime. The 230 climbers used to build out the first ribbon only need to last for the few weeks of their operation, whereas the working 20 ton climbers might want a more long lasting device.

If the Compower FEL system is selected, specifically designed GaAs photovoltaic cells can be used with 59% conversion efficiency, 82% filling factor and a sizzling power density of 540 kW/m^2 [D'Amato, 1992, and Charlie Chu at Tecstar, private communication].   This is very impressive. For those of you who are familiar with solar power you may recall that we will do quite well to get half a kW/m^2 out of solar panels. That is because as big as the Sun is it is still a diffuse light source, and the power comes in at all wavelengths, most of it wasted because PV cells can only convert a certain range of frequencies to electricity.   Lasers are a far more concentrated source and produce their light all at one wavelength, which can be absorbed by specifically designed PV cells with high efficiency. The smaller climbers with a 4m diameter base has 12 m^2 of area and needs to produce 100 kW of power.   That only requires 8.3 kW per square meter of PV cells, an easily achieved capacity given the generous margin from their 540 kW/m^2 potential. These cells would also work well with a large laser diode array as proposed by Kwon, 1997.

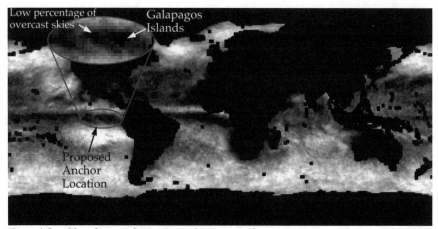

Fig. 4.2: Cloudiness frequency from satellite microwave data. The dark areas show regions of low frequency of clouds.

One additional problem that we need to address is lost transmitting time because of overcast skies. At our proposed anchor location where it would be best to also place the power beaming facility, the percentage of overcast skies appears to be low (figure 4.2) but to insure continuous operations a second beaming facility located in a separate weather zone would be advisable. It would be extremely embarrassing, in this highest of high tech transportation systems, to have to tell one's passengers that they are out of power because of overcast sky's at the landing site. In our proposed situation the second beaming facility could also be located on a movable ocean platform (see Chapter 6: Anchor) roughly hundreds to thousands of kilometers from the anchor or in the mountains of Ecuador (10,000 ft altitude). An additional power beaming system in the United States (Mojave desert [Bennett, 2000]) could also be used for supplying power to climbers above 10,000km.

Microwave Power Beaming

Several studies have been conducted on the beaming of power from space using microwaves [Brown, 1992: Glaser, 1992]. Primarily for use with large solar panel power stations in geo-synchronous orbit to beam the power to receiver fields on earth where it would be converted to electrical power. These studies have looked at frequencies of 2.4, 35 and 94 GHz primarily and utilize dish, flat or phased array transmitting and receiving antenna [Brown, 1992: Koert, 1992]. If we consider our specific situation of beaming power to space and not from space in these same terms we start with the equation:

$$\frac{P_r}{P_t} = \frac{A_r A_t}{d^2 \lambda^2}$$

where P_r is the power received, P_t is the power transmitted, A_r is the area of the receiving antenna, A_t is the area of the transmitting antenna, d is the distance between the transmitting and receiving antenna and λ is the wavelength. A low-mass receiving antenna is required so we will select a baseline 3 meter diameter area (A_r =7m², 30kg). We also need 50 kW delivered to an altitude of 15,000km (for the initial climber, 40 times this for the final climbers). To deliver this power to our receiver we will need a phased array transmitting antenna of at least 1×10^6 m² (1km²), not something that is easily deployed on the open ocean. Including rectenna (rectifying antenna) efficiency (50% [Koert, 1992: Koert, 1999]) and transmission efficiency (30% [Koert, 1992]) we find we will need 1.7×10^5 MW, or 792 MW, or 110 MW, going to the transmitters for 2.4 (λ= 12.5 cm), 35 (λ= 8.6 mm), and 94 (λ= 3.2 mm) GHz respectively for the first climbers. This system is expandable as required by putting more power through the phased array and the received power is proportional to the transmitting antenna area. This is a little outside our budget range; the 2.4 GHz (λ= 12.5 cm) station, using 1.7×10^5 MW of power, or 170 GW, is about one quarter of the entire USA electrical generating capacity; undoubtedly California would object. A frequency of 94 GHz is definitely preferable from the numbers above. Considerable effort has gone into developing rectifying antenna at 35GHz for use as lightweight receivers. These rectennas have 50% total efficiency and similar results should be achievable at 94 GHz [Koert, 1999]. The mass of a rectenna would be comparable to lightweight solar panels at 33kg for a 50 kW receiver [Koert, 1999].

Microwaves at frequencies above 10 GHz are readily absorbed by water vapor (easily 50% absorption at 94 GHz) so careful high-altitude or dry site selection is required. If we go to the longer wavelengths where absorption is less of a problem we find the efficiency of the system drops dramatically unless a very large transmitter (1600km²) can be built, a difficult proposition to say the least.

Power Beaming Summary

In examining the two possible systems we see that there are performance, maturity and operational differences. An overall summary of the two systems is shown in table 4.1. It is clear that at this time the laser power beaming system is the better choice of the two. The higher efficiency, smaller transmitter, and maturity of the laser beaming system are all distinct advantages over the microwave system. These same characteristics would also be reflected in a substantial cost difference as well. The cost of microwave facilities would be significantly higher than a laser system and the ongoing daily cost of power input of the microwave system would negate most of the advantages of a space elevator. Construction of the Compower system would eliminate any concerns on the construction or performance of the power beaming system.

Table 4.1: Laser vs. Microwave Power Beaming

	Laser	Microwave
Operating wavelength	0.84 microns	3.2 mm (94 GHz)
Transmitter System	Free-Electron laser / deformable mirror	Phased array
Transmitter area	12m diameter	1km diameter
Receiver system	Tuned solar cells	Rectennas
Overall system efficiency	3% - 14%	0.05%
Power input levels	Within reason	Outrageous
High altitude operation	Preferred	Preferred
Development level	Under Construction	Design stage

While we've got it, use it

As mentioned in chapter 3 we were dissatisfied with the original plan to go with liquid fueled rockets for the trip from LEO to GEO and decided to develop a design using electric powered engines. One of the main factors in this was the realization that we already had our power system. Here we are working to adapt the Compower laser system to power our climbers when in fact the Compower system was originally developed to power satellites in orbit from the ground. Some of the work on an aiming and tracking system has already been done. The same lasers that we intend for the climbers are more than powerful enough to power the spacecraft. And since they must be completed and tested before launching the ribbon and they will be mounted on ocean-going platforms, they can be moved to strategic points around the world to power the spacecraft. During the last two months of the orbital maneuvers only one station will be needed and that aimed at a fixed point in the sky. This leaves plenty of time for the furthest stations to be re-gathered at the anchorage site to resume business with the climbers. As to the spacecraft, the same type of photocell panels developed for the climbers should do nicely, in this case about 10 meters in diameter.

The next chapter goes into more detail of just how this power and tracking system works.

Note 1:
The long-term beam radius can be expressed as:

$$\langle r_L^2 \rangle = r_d^2 + \frac{4L^2}{(kr_o)^2}$$

where r_o is the transverse coherence length, L is the distance from the transmitter to the receiver, D is the diameter of the transmitter, the diffraction limited spot radius is

$$r_d = \sqrt{\frac{4L^2}{(kD)^2} + \left(\frac{D}{2}\right)^2}$$

$$k = \frac{2\pi}{\lambda} \quad \text{and}$$

$$r_o = \left[1.46k^2 \cdot \sec(\phi)\int_0^L C_n^2(\eta)\left(1 - \frac{\eta}{L}\right)^{\frac{5}{3}} d\eta\right]^{-\frac{3}{5}} \qquad \text{(the transverse coherence}$$

length)

In the case where $\eta \ll L$ when $C_\eta > 0$

$$r_o = \left[1.46k^2 \cdot \sec(\phi)\int_0^L C_n^2(\eta) d\eta\right]^{-\frac{3}{5}}$$

Following the example in the book we use the stated CLEAR I night model (this model may have some problems in our situation because the model starts at 1.2km altitude):

$$r_o = 2.76cm \quad \text{for} \quad \lambda = 0.5\mu m$$

so for a climber at 10,000km, a 10m transmitter and 0.5 micron wavelength we get:

$r_d = 6m$

then

$$\langle r_L^2 \rangle = 6^2 + \frac{4 \cdot 10,000,000^2}{\left(\dfrac{2 \cdot \pi}{0.5 \times 10^{-6}} \cdot 0.0276 \right)^2}$$

$r_L = 58m$

This long-term broadening can be separated into two components.

$$\langle r_L^2 \rangle = \langle r_S^2 \rangle + \langle r_C^2 \rangle$$

where $\langle r_S^2 \rangle$ is the short-term beam broadening and $\langle r_C^2 \rangle$ is the beam wander (drift).

For the short-term beam broadening we have:

$$\langle r_S^2 \rangle = r_d^2 + \frac{4L^2}{(kr_o)^2} \left[1 - 0.62 \left(\frac{r_o}{D} \right)^{\frac{1}{3}} \right]^{\frac{6}{5}}$$

$r_S = 57m$

For beam wander we have:

$$\rangle r_C^2 \langle = 2.97 \frac{L^2}{\left(k^2 r_o^{\frac{5}{3}} D^{\frac{1}{3}} \right)}$$

$r_c = 9.3m$

Chapter 5
Deploying the Initial Ribbon

We have two problems to consider when deploying the ribbon. Once on station in geo-sync orbit we can start spooling out the ribbon by applying power to the ribbon deployment yoke attached to the axle of the spool. But without some force to pull the ribbon outward the ribbon would just pile up like a plate of spaghetti, or rather a floating plate of spaghetti. Secondly, once a length of ribbon was deployed there is nothing to tell it which direction it should be going. This is not a new problem in spacemanship, any satellite that needs to point at the Earth to do its job must rotate once per orbit, this is the spin-rate; guidance and station keeping engines are designed to do this. But in our case we are going to change the shape and center of gravity of the spacecraft, or at least its attached components.

Therefore we need to impart angular momentum to the ribbon as it is deployed to insure proper orientation. If not controlled the ribbon would eventually orient itself vertically due to gravitational gradient forces but the loose end could just as well point away from Earth as toward it. Some force must be input to insure proper alignment. The most critical point is early in the deployment process, once the orientation is well established, with a sufficient length of ribbon, the gravitational gradient forces will keep the ribbon vertically aligned. In addition, when these gravitational forces start to show a significant influence, (early in the deployment) we must be prepared to dampen any oscillations that may occur.

Initially we will deploy the ribbon with a mass on the end, enough mass to be dominant over the mass of the first kilometer of ribbon. We will make this mass a mini-spacecraft with a low thrust, long burning engine and the guidance capabilities to accelerate itself in a specified direction. We can thus deploy a length of ribbon and give it an angular momentum such that at least initially it is in the proper orientation.

Angular velocity is: $\omega = 2\pi/\text{Period}$ radians per second. This, times the radius (the length of ribbon) gives the velocity needed for the endmass. If we deploy 1000m of ribbon and want it to rotate once per day then the endmass will be moving at 0.073m/s. This velocity is an easy requirement to meet; it can be imparted to our endmass with a very small, standard monopropellant system.

The next question is how do you deploy the ribbon further. Here we will take advantage of the variation in gravity across the length of the ribbon. This difference in gravity will pull on the ribbon such to keep it pointing at Earth. This is called a gravity gradient torque. For this first example let's use end masses of 10kg and an initial deployed length of 1km. In this case the effect of the ribbon mass (0.1kg) is very small, its contribution is only about 1% of the total moment, and integrating over the 1000 meters is messy so we will ignore it for the moment to show the effect of the end masses. So, the moment of inertia will be:

$$I = mr^2 = 10kg \cdot (1000m)^2 = 1 \times 10^7 kgm^2$$ for the ribbon with end mass.

The angular momentum will be

$$L = I\omega$$

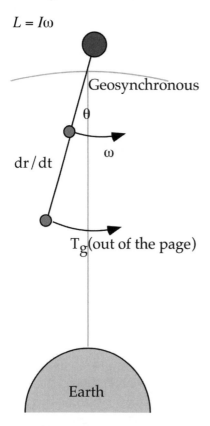

Figure 5.1: Gravity gradient diagram

Now if we extend the ribbon to 2km the moment of inertia increases, the angular momentum will remain constant so the angular velocity decreases and the ribbon will begin to drift out of its proper orientation. As soon as it leaves a vertical orientation it will feel a torque due to the gravity gradient (figure 5.1) described by:

$$T_g = \frac{3\mu}{R^3}\left|I_z - I_y\right|\theta \qquad \text{(\textit{Space Mission Analysis and Design}, 1991)}$$

$I_z \gg I_y$ so we will use I for I_z, so we can simplify, $\qquad T_g = \frac{3\mu}{R^3}I\theta$

where T_g is the gravitational torque, R is the orbital radius, I is the moment of inertia, μ is Earth's gravity constant ($3.988 \times 10^5 km^3/s^2$), and θ is maximum deviation (swing) of the Z axis from vertical.

What we want to do is deploy the ribbon (increasing the moment of inertia) while keeping the angular velocity constant. To do this we need the spin angular momentum to increase at the proper rate. The rate at which the gravity gradient torque imparts angular momentum to our ribbon is dependent on the moment of inertia and angle of the ribbon to vertical. What this means is that if we deploy the ribbon at the right velocity and at the right angle to vertical the gravitational gradient torque will keep the ribbon at the same angular velocity and in the same and proper orientation.

From the above formula we can derive the maximum deployment rate which turns out to be extremely fast. An angular deviation from vertical of 10 degree gives us a 3-day deployment rate to the Earth and a 1-nanosecond deployment to two kilometers. (A moments thought will show that this last is faster than the speed of light so we shouldn't have a problem going too fast.) It implies that once the ribbon is orientated and deployed to more than a kilometer that it is stable and deployment can go at any physically conceivable rate, the gravitational gradient torque will keep the ribbon vertical.

One of the things that we have ignored here is the mass of the ribbon. During the initial deployment (the first kilometer) the mass of the ribbon is small compared to the endweight. Once enough ribbon is deployed so that the ribbon mass is comparable to the mass of the endweight (about 100km) the ribbon orientation will be fixed. Eventually the mass of the ribbon dominates the moment of inertia but the difference is a constant and appears on both sides of our equation so the deployment rate is unaffected. This also implies that our overall energy requirement (fuel) for this maneuverable endweight is rather modest.

The second thing that we have ignored in the calculation above is conservation of angular momentum. If we impart spin angular momentum to the deploying ribbon, in a closed system, angular momentum has to decrease somewhere else. In our case the Earth's gravity and rotational acceleration are pulling on our ribbon. Since the ribbon is sometimes at a slight angle relative to vertical, part of the gravity gradient forces are converted into a force along the ribbon resulting in forces along the spacecraft's orbit instead of perpendicular to it. Due to our particular situation where the ribbon is comparable in size to the gravity well, we have a net force on the spacecraft and ribbon that reduces the orbital velocity; we put drag on the deploying spacecraft. When this happens the ribbon is given spin angular momentum and the spacecraft will lose orbital angular momentum and will want to drop to a lower altitude.

A second fact that complicates our calculations is that as the ribbon is deployed different parts of it experience different gravitational acceleration. This changes our apparent mass distribution and if we want to maintain the center of this mass at a geosynchronous orbit all during deployment we must dramatically increase our orbital angular momentum by continuing to apply thrust to the deploying spacecraft. In our specific situation the geosynchronous orbit altitude for our center of mass depends on how much ribbon we have deployed. Suddenly we have stepped into a fairly complex situation where we must consider much more than just the ribbon.

Getting the system in orbit and deployment

Let's examine our deployment situation in a little more detail. First, let's say that we want our final orbit to be geosynchronous, that is, have a 24-hour orbital period. This is required for realistic use of the space elevator. Second, we will say that once the end of the ribbon touches Earth that no more angular momentum will be required to be supplied on-orbit. Once the end of the ribbon is anchored all of the angular momentum that will be supplied to spin up the spacecraft and ribbon will come from the Earth through tension in the ribbon. Prior to the ribbon touching down we will need to supply the required orbital angular momentum with the use of propulsion systems on the deploying spacecraft. First we will need to get the ribbon, spacecraft and propulsion system into geosynchronous orbit (the spacecraft and propulsion system will be counterweight material later). The last thing that needs to get folded in is we need to do this with reasonably priced systems.

Let's begin by laying the groundwork. After examining the advantages and disadvantages of various ribbon lengths, a length of 100,000km was decided. This will allow us to reach the inner solar system, (details later) and gives us a favorable ribbon-mass to counterweight-mass ratio for this length of 1.364 (see Chapter 2). Now we only need to have the mass for one other item and we will be able to define (from the above discussion) the approximate mass of all the other components. That one item is the mass of the initial ribbon we will be sending up.

Table 5.1: Some of the Launch Vehicles to be Considered

Vehicle	Payload to LEO (kg)	Development level	Development cost	Launch cost
Delta IV Heavy	23,040	Operational	NA	Est. $200M
Atlas V	18,000	Operational	NA	$90M
Shuttle	24,400 (204km)	Operational	$10B in 1977	$245M (1988) $500M now
Shuttle-C	77,000 (400km)	Design stage (3 years required)	$3B-$7B	Comparable to Shuttle

Table 5.2: Upper Stage propulsion Systems

Upper Stage	Isp	Thrust (N)	Comment
Centaur	446	147,000	Cryogenic (~$30m)
Star 48	206	66,300	Solid
Ion prop	2000-6000	5e-6 – 0.5	Large dry mass
MPD*	2000-4000	25-200	Space proven
MMH / NTO**	325	503	Qualification testing

NOTES:

The shaded information denotes areas that could be a problem for the propulsion system in our application.

*Magnetoplasmadynamic

**Information from TRW, private communication.

All other information from: Space Mission Analysis and Design, 1991

The propulsion systems and launchers we can utilize for our first attempt at building a spaceship include some currently available ones and possibly ones that would be developed for this program (table 5.1).

Our first objective is to deploy as large an initial ribbon as we can, given other tradeoffs. (The main reasons, which we will go into detail later, are the survivability of a small ribbon and the time and effort of the later build-up.) We could for example send up many spools of ribbon segments and splice them together for a rather large starting ribbon. We looked at this, and after the list of problems got long enough we next decided to try it with the largest single ribbon we could lift. Allowing for mounting hardware, this would be a 22,000kg ribbon on the Shuttle. (This was before one of the other systems, above, became available.)

We can use standard shuttles or large expendables to get our payload into low-Earth-orbit. No matter what system we put together the complete system is really big and we can't hope to place the entire system in geosynchronous orbit directly from Earth. Accepting this then our best option will be to place the parts in LEO, assemble them there with the assistance of the space station or shuttle crew and then lift the completed system to geosynchronous. This will avoid on-orbit ribbon splicing since the ribbon can be sent up as one piece and then attached to the remaining components. After the pieces have been assembled in LEO we will need a second stage spacecraft to get us to GEO. Table 5.2 has information for both LEO to GEO stages and on-board propulsion systems. The basic scenario for getting our initial ribbon deployed is laid out in figure 5.2.

Given the notorious cost of building anything new for space, the obvious first approach is to use readily available components. As mentioned in chapter 3 if we use conventional methods it would then take a cluster of several (7) liquid fueled Centaur rockets to get to GEO and several of the low thrust units (like the MMH/NTO system) to impart the required angular momentum as we deploy the ribbon and move outward from GEO. Shuttles or the large expendables are appropriate for several reasons: 1) They exist. No development is required. 2) Our system can be disassembled into components that will each fit on these launchers. 3) The launchers' capacity allows us to deploy the space elevator in a reasonable number of launches.

Table 5.3: Calculated Masses for the off-the-shelf Deployment System

Ribbon mass	22 tons
Spacecraft mass	5 tons
MMH/NTO system	2 tons
MMH/NTO propellant	18 tons
Centaur system	12 tons
Centaur fuel	127 tons
Total mass of payload placed in LEO	186 tons

Using a set of mass ratio equations we approximated the masses of the components for this version of our spacecraft system (table 5.3) and found we did indeed have a massive project on our hands.

Notice that 145 tons of that is fuel! And it would take 8 to 9 launches to get this all up to LEO. The launch cost alone would come to about $4 billion dollars. Now this is doable and worth going ahead if necessary, but it is cumbersome, expensive and has several problems, for instance storing the cryogenic fuels while waiting for everything to get launched from Earth and assembled in LEO. The search was on for a more effective way.

The Shuttle-C would have several advantages. It would cut the number of launches to three and allow for a 20% larger initial ribbon to be placed in orbit. But unless its $3B to $7B development costs were being covered by some other program we would be well behind to take on these cost under our program. But there is progress on other fronts. Two recent developments may save the day.

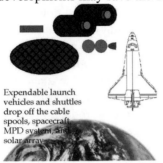

Fig.   5.2: Deployment scenario for the   initial ribbon.

Expendable launch vehicles and shuttles drop off the cable spools, spacecraft MPD system, and solar array

Parts are assembled in LEO

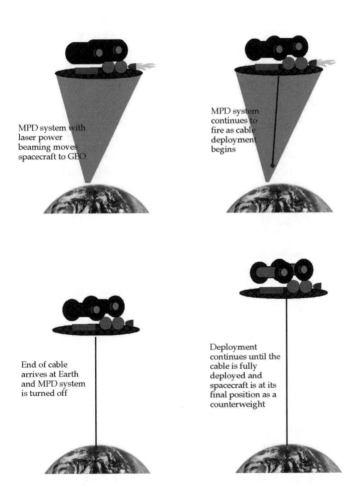

MPD system with laser power beaming moves spacecraft to GEO

MPD system continues to fire as cable deployment begins

End of cable arrives at Earth and MPD system is turned off

Deployment continues until the cable is fully deployed and spacecraft is at its final position as a counterweight

The Delta rocket is being upgraded to accommodate larger loads, now designated the Delta IV Heavy. As noted in table 5.1 it now has a 23,000kg capacity to LEO, allowing us to avoid having to put the ribbon spool on the more expensive Shuttle and also reducing the total number of launches. The second item to come into focus is the MPD (Magneto-plasma-dynamic) system. The research has progressed to the stage that good data is now available for operating size units (as opposed to small test-bed units) and the expected costs of pushing on to a production model are much lower than thought previously. It now looks like *new* may actually cost less than off-the-shelf. This led us to look at substituting an MPD system for the entire high thrust, and fuel heavy, LEO to GEO trip and also for the post GEO ribbon deployment maneuvers.

Once we had decided on laser beaming as the method of delivering power to the MPD system everything else started to fall into place. When we ran the mass numbers for a scenario using the MPD system as LEO to GEO transport for a 22 ton ribbon spool it only came to about 44 tons departing LEO! (By comparison, the liquid fuel rocket model would have been from 168 tons at best up to 186 tons.) With the first objective being to launch the largest initial ribbon possible it was clear from these mass ratios that we could afford the mass budget to launch two ribbons spools if we could figure a way around the difficult problem of splicing ribbon on orbit. The solution for that was, don't splice on orbit, just tie the ends together and deploy both ribbons side by side. After deployment and anchoring (the deployment problems are now over) use the first ribbon-dispensing climber, whose splicing job it is anyway, to zip the two halves together with a strand of added ribbon. Don't you just love it when a plan comes together?

Deciding on two spools of ribbon, 20 tons each, (being conservative, don't push the limits so much) we now had all the items we need to develop a comprehensive model for our spacecraft and its deployment scenario.

The Delta V approach

From basic spacecraft design we can set out the approximate mass of the components. Reasonably close is good enough at this stage of the process, the actual design studies will take some serious money, and what is most important is to get a good idea of the fuel requirements and the overall mass at each stage of the journey.

Let's work the problem backwards. We know that in conventional high thrust rocketry there are four trajectory changes, or rocket firing sequences, to deploy our ribbon. The first three maneuvers are to get into GEO, the fourth is the ribbon deployment maneuver as the spacecraft is moved from GEO out to its end point. For each of these four maneuvers we can compute the delta V (ΔV) and given the mass to be moved, the fuel required.

Taking them in reverse order the first maneuver to compute is the last one, the deployment from GEO to end point. This maneuver is different than all the others that experience a change of angular velocity. For this we can calculate the angular momentum when we first get to GEO and again when the ribbon touches the ground. This difference can be supplied by the MPD system we are already using since it is a low thrust system able to fire continuously. The final angular momentum we will need to do numerically (calculate all the parts on a computer). The reason for this is that our ribbon is not a simple shape. To do this we simply calculate the moment of inertia for each small part of the ribbon and spacecraft and add them together. Since the angular velocity is the same for the entire system we can multiply the moment of inertia and the angular velocity to get the final angular momentum. The difference between the initial and final angular momentum will be equal to the torque supplied by the MPD system.

As the ribbon is deployed the MPD system, all its fuel, and the remaining ribbon on the spool is acting as counter weight. The MPD system is driving this package away from Earth while its mass decreases with the expenditure of fuel and ribbon. The spacecraft will be at 63,170km altitude when the ribbon just touches down on Earth, at which point there is little fuel left and the rest of the ribbon is deployed by centripetal force, the result of being tied down on Earth. Being tied down to a rotating Earth also imparts the rest of the angular momentum as the ribbon deploys to full length; the Earth literally drags the ribbon around until it is fully fixed, vertically.

We would prefer a burn time close to the time it takes to deploy the ribbon to Earth to minimize the forces on the spacecraft and ribbon, so the MPD system will probably be slightly different than what we are using here. (We will find that if we use four weeks the thrust of the MPD system should be about 30 N, which fits well with this type of system.) This maneuver is estimated to require approximately 1.1km/s ΔV. We will set this answer aside for the moment while we work on the three maneuvers from LEO to GEO.

LEO to GEO the fast way

We know that there are three trajectory changes to get into GEO, and we want to compute their ΔV's. The first burn is to leave LEO with a geocentric transfer orbit (GTO) insertion. This changes the orbit to a wide ellipse, raising the apogee to the new orbit, 35,785km, with the perigee at the starting orbit, for instance, 300km. So far we are just visiting GEO, this is a wide elliptical orbit that will come back down to 300km. To restate this so that the picture is clear: An elliptical orbit has a high point and a low point, relative to Earth in this case. The speed of a spacecraft at the low point is faster than the circular orbital speed for that altitude and the high point has a speed slower than the circular orbital speed for that altitude.   After the first burn, the GTO insertion, we will rise, momentarily, to GEO but be slower than circular GEO.

The second trajectory change (burn) is to change planes. If we launch from Cape Canaveral, as we must do if we use the shuttle or other heavy lift vehicle for one or more of the launches, the parts of our package will accumulate in an orbit inclined 28.5 degrees to the equator, e.g. the latitude of Cape Canaveral.   To be in GEO we must bring the spacecraft into the equatorial plane, thus we must perform a plane change maneuver. The formula for this involves a velocity factor; $2V*\sin(\theta/2)$ where θ is the angle of change. So, it makes sense to do the maneuver when the ship is at its slowest velocity, which is just at the top of the ellipse, at apogee; before(!) we "circularize" to stay at the new higher orbit, which is a higher speed. The third trajectory change is to circularize the orbit, to raise the perigee from 300km to GEO, 35,785km and thus stay in GEO. This firing also needs to be performed at the top of the ellipse, to speed up to the orbital velocity for GEO.   The second and third trajectory changes are computed individually but can be done at the same time.

The "two burn" method of changing orbits (not counting the plane change for the moment), with one burn at apogee and the other at perigee is right out of the ellipse equation, (see addendum), and gives the velocity changes needed to follow the most efficient orbit, a Hohmann orbit, to a new position. It works just fine for high thrust liquid fueled rockets because the new velocity, which puts the spacecraft onto the Hohmann pathway, comes immediately, it only takes a couple of minutes for the burn. Here is what the trip to GEO looks like using the, quick and easy, ellipse equation for a first approximation.

Delta V to GEO

For high thrust, liquid fuel engines, starting from 300km LEO.

First burn, GTO insertion:	ΔV 2.43km/s
Second burn, plane change:	ΔV 0.79km/s
Third burn, circularization:	ΔV 1.47km/s
Total to GEO	ΔV 4.69km/s
Fourth "burn", deployment	ΔV 1.10km/s
Total mission	ΔV 5.79km/s

Amazing as it seems, this costs more than going to Mars and we haven't even left the neighborhood.

Now all we need are the other masses that are going to be making the trip in order to compute the fuel load. The ribbon we have set at 40 tons, we can get the other masses from what we already know, and the general nature of spacecraft components. With a mass estimate we compute the fuel needed for the delta V above. Here is a first approximation of the mass distribution.

Table 5.4: Masses Calculated from desired limits and required ΔV

Ribbon mass	40.0 tons
Spacecraft mass & Deployment hardware	10.0 tons
Spacecraft frame, C&C, power panels, etc.	3.0 tons
MPD Engine system	2.0 tons
Fuel tanks	2.0 tons
Additional Counter Weight mass	12.4 tons
Dry mass leaving LEO	69.4 tons
MPD fuel	11.0 tons
Total mass placed in LEO (Start Mass)	80.4 tons

These are the same figures, abbreviated, down to "Dry mass leaving LEO" that we presented in chapter three but the fuel mass and therefore the total mass is lower. That is because, except for the fourth burn of ΔV 1.1, the above delta V's are wrong for low thrust engines.

Orbital mechanics for slow burning engines

First let's acknowledge that the above fuel mass is close, it's in the right ball park, we know that this configuration is going to be a little over 80 tons and we could do general mission planning on this. It's enough to set a major mission parameter; it commits us to using 4 of the Delta IV Heavy brutes for launch with many tons to spare before having to go to five. But there is an important, and as it turns out a very interesting, engineering factor involved here and having a more exact figure is going to influence any later changes that might be made to the program.

To illustrate the problem we will make a simpler example with the focus on just the main orbital change from LEO to GEO, and leave aside for the moment the plane change and the fourth burn ribbon maneuvers, but we will include the fuel mass that they would represent. This will give us consistent numbers to use in another explanation coming later.

The conditions: MPD engine, Isp 4000, power (P) 800 kW at 75% efficiency (e), that gives an exhaust velocity (Ve) of 39,200 m/s, a thrust (T) of 30.6 Newtons, and a fuel flow rate (Fr) of 0.780925 g/s. We can simplify the masses to just 3 items. All the spacecraft components and the payload, 57 tons, the extra counterweight mass, 12.4 tons, and the fuel for the later maneuvers, 4.78 tons for a total "end mass" at the end of the GEO leg of the trip of 74.2 tons. Now all we need is the fuel for this leg to have the total mass to start the trip.

The fuel is determined by the ΔV, the velocity change, of the trip, and the normal way to figure ΔV is to use the ellipse equation and the "two burn" method at the bottom and top of the ellipse as mentioned above in "Delta V to GEO". As we noted this would give us two trip segments of 2.43km/s and 1.47km/s for a total of 3.9km/s ΔV. (By the way, this would put us at GEO altitude, but without the plane change burn we would not yet be in a GEO orbit.) But, this method of calculating momentum changes only works under specific conditions.

The assumptions in the two burn ΔV calculation are:

1 - that the first burn injects you into a Hohmann orbit, the most efficient, minimum energy path, to the next orbit altitude.

2 – that this burn happens instantaneously, or at least fast enough that no one notices, e.g. that the change in condition from velocity 1 to Velocity 2 is not spoiled by extraneous forces like gravity working for some time.

3 - that the circularizing burn at the other end is also fast enough that its change of condition is not effected.

None of these conditions exist in the slow burn ΔV situation.

With the continuous burn of low thrust engines we are exerting millions of mini orbit transition trajectories. Each one may in fact be a Hohmann (H) orbit and therefore the differential ellipse equation does apply to each one, even averaging a cluster of them, but multiple small H orbits do not add up to one efficient H path. We are not taking an efficient path to higher orbit, so we can't use a one orbit (actually 1/2 orbit) minimum energy answer; additional mathematics will have to be applied. A search of the literature for the missing elements proved less than satisfactory.   There were methods using highly complex statistical simulations that took several hours to run to a solution, and then only an approximation; one with a 5% error factor. A program from CalTec stated the problem boldly, "Unlike impulsive thrust trajectories such as the Hohmann trajectory, this class of maneuvers cannot be described by analytical equations." There you have it; you can't get there from here.

In unscientific terminology, this is silly. Maybe there isn't an "analytical equation" but there has to be a way to model the conditions, in reasonably abbreviated form, such that a meaningful estimate can be made, and Right Now. What we want to do is to integrate the increments of orbital change from LEO to GEO under constant thrust. Since we have it on good authority that there is no integral we will just make one by setting up a stepwise table on the spreadsheet. For a first try we took orbital increases at what was admittedly a rather coarse interval of 100km, starting at 300km altitude, (i.e. 300, 400, 500 etc.) and using the rather long two step ellipse equation computed the ΔV for each segment. This seemed to work fine so we finished out the table and summed the column of tiny ΔV's. The answer was 4.652km/s ΔV, 20% higher than the 3.9km/s of the pure H orbit. This looked about right for the amount of inefficiency but how do you check it out and also how much error is in the segment interval of 100km?

In a paper on ion propulsion, Keaton suggested that the delta V for these engines was the difference between the orbital velocities. Now subtracting orbital velocities is just not the way to do it. It shouldn't give you anything meaningful; the difference between Earth orbital V and Mars orbital V is 5.7km/s but if you used that much speed in a rocket ship trip to Mars you would badly over shoot. But, as it turns out, he is right! In our case the difference between the orbital velocity at 300km and 35,785km is 4.653km/s, and the sum of our column of small ΔV's is 4.652!? How could there be such a coincidence with the two values? Meanwhile, back at the 100km segment intervals, we checked the ΔV value both ways, and the difference between the ellipse value and the "subtraction method" value was 7×10^{-7} at 400km and 5×10^{-10} at high obits (34,000km). This is nothing! They give the same answer! The subtraction method is a direct representation of the ellipse equation over small intervals! And summing those intervals gives the difference between the velocities of the starting orbit and the ending orbit; you don't have to do the stepwise table!

But is this the correct velocity expenditure (ΔV) equation for determining fuel usage for slow burn engines? This may take some more rigorous mathematical proof, but it does appear that the subtraction (net difference) method is valid. By taking the difference between the two circular orbital velocities you are representing the amount of INefficiency, the excess, over the H pathway. Or, the subtraction of orbital velocity differences represents the real world inefficient path of slow orbital transitions. And that is exactly what we are looking for, a way to measure the inefficient pathway. Therefore, for continuous velocity changes, (slow burn) a summation of small net velocity differences should equal the gross net difference between the trip starting and ending velocities, within the error of the stepwise math. And it does!

Plugging 4.652 into the delta V equation gives us 9.337 tons of fuel for this portion of the trip, 14.12 tons for the whole mission, bringing our total mass to 83.46 tons. Now that we know the amount of fuel we can divide by the rate the fuel is burned to find out how long this slow boat will take to get to GEO. Tons to grams is one million so 9.337 million grams divided by 0.78093 grams of fuel per second, equals ~12 million seconds or 138.39 days to GEO and many more days for the rest of the mission. And that's if the engines burn continuously, which they won't.

Where angles fear to tread?

Or angels either, for that matter. Now planning really gets sticky. With the power for our electric engine coming from a laser beaming station on the ground we have to know how long the satellite will be "in view" of the station as it whips by overhead. We just figured that it will take 138.4 days of engine burn time to get to GEO, but the engines will only be on for the few minutes of each orbit that are in the beam. So, how many total days is it really going to take? We could put up a dozen stations around the world and be sure that the power would be on all the time, but we haven't got the budget for that. So, let's work the problem for one station, get a worse case condition, and then see how much we can cut down the time with the two or three stations that we could afford.

The starting orbit is LEO at 300km. All the math is going to involve the total "R" length from the center of the Earth to the orbit altitude. So while we refer to the orbits in the convenient fashion of 300, 400, 500km etc. each step will have the distance from the center of the Earth to the surface (6378km) added on. From the ellipse equation [short version for circles, $(GMe/R)^{1/2}$, see addendum] we get an orbital velocity of 7.7273km/s for the 300km circular orbit. Take the circumference of the orbit, 41959.11km and divide by the orbital velocity, to get the time of one orbit, 90.499 minutes. That is one orbit relative to the center of the Earth regardless of what the Earth is doing in the meantime; call it the *absolute* orbital time. Since the Earth is doing something in the meantime, the orbital time from our point of view, say from straight overhead around to straight overhead again, is longer by the amount of time the Earth turns during the satellite's orbit. Both the satellite and the Earth are turning in the same direction so it is an overtaking problem, which is, the whole distance divided by the difference in the speeds. In this case the whole distance is one circle or 360 degrees and the difference in the speeds is 3.9779 degrees/minute for the satellite, minus 0.25071 degrees/min for Earth or 360/3.72722 = 96.587 minutes; call this the *relative* orbital time. That is, the orbital time relative to an observer or a beaming station on the Earth. Note: we use sidereal day of 23.93183 hours for better accuracy, (it will be noticeable, where we are going with this).

Now we need to find out how much of this orbital time the satellite will be in view of the beam. Assume that the station can track the full 180 degrees from horizon to horizon. We can set up a right triangle and solve for two points on the orbit, the starting point on the horizon where the satellite comes into "view" (point a) and the ending point on the opposite horizon, and measure the degrees of arc between those two points, see figure 5.3. We know r, 6378km, and for the lowest, starting orbit, we know h, 6378 + 300, so we can find angle A = 17.24 degrees and double that for the entire viewing arc of 34.48 degrees. This is 9.58% of the relative orbital time of 96.59 minutes so the spacecraft gets 9.25 minutes of power beaming time for the first orbit from one station. This angle gets wider as orbital altitude increases (point b) and each orbit gets longer in actual time as well as beaming time from one station. (120 degrees at 6,400km.)

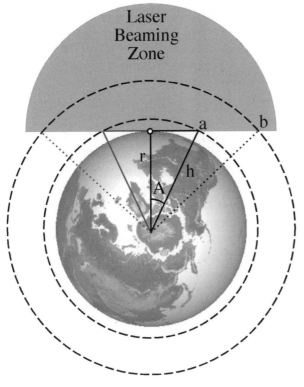

Fig. 5.3: Angle of "View" for power beaming.

Since we know the fuel flow rate   we can figure the new velocity and from that you can figure the next higher orbit that would result from each burn, and in turn the view time and burn time for the next orbit. When the last orbit gets to GEO, stop and add up all the orbits and time it took to get there.   We can make this easy with a spreadsheet. Take the orbits in clusters of 100km altitude increase, (i.e. from 300km to 400km) and average the orbital and viewing (burn) times.   Combine with the velocity change (from 300km to 400km) and the fuel used, and relate that to the total trip to GEO of 4.652 ΔV to get 16.54 days for the first leg to 400km. You can tell this is going to be a really slooow trip. The total of all the 100km segments comes to 481 days using one station, (just LEO to GEO leg).

Now we need to bring back into the discussion the other two burns.   The plane change maneuver that we originally computed as 0.79km/s ΔV at the top, (the slowest part) of an elliptical transfer, also does not apply to slow burn engines. We never get that slow; the slowest velocity will be, as we get to GEO, 3.07km/s, making it a 1.51km/s ΔV maneuver for the given angle of change.   And we can't save time by combining it with the LEO to GEO trip, we would just subtract from the energy or time devoted to increasing the altitude.   Therefore this leg will be 42 days and about 2.8 tons of fuel for an Isp 4000/s engine. The ribbon deployment to end of mission ΔV that we computed as 1.1km/s ΔV stays the same and would consume 29 days and about 2 tons of fuel for the same engine. This is 71 extra days, bring the total one-station time to 552 days. Ouch!

Let's compromise

This is much too long, it is obvious that we need at least three beaming stations, which would cut the 481 days down to 172 days (not 1/3 of 481; the station "views" overlap above 6400km). And that is not all, after we get to GEO, we still have the plane change maneuver of 42 days and the ribbon deployment maneuver of 29 days, so it is still a long 235 days. What else can we do to shorten the time? "Throw the canary another seed." Placing more stations around the world would help but only a little. At altitudes below 6400km three stations do not provide complete coverage so four would only help up to there (but not past), decreasing the time to GEO to 152 days, (not to 121), plus upper maneuvers = 223 days. But above 2700km more stations won't help because now the four stations start overlapping, so you no longer get the full multi-station effect (it's just full time power).   With the power beaming system the spacecraft is only "in view" of one power beaming station for as little as 10 minutes out of each orbit, and that is the only time it would be under thrusting power. Since these lowest orbits are the most troublesome, with short exposure times, the only other thing we can do is use a booster engine system and try to get the first 1000 to 2000km increase out of the way quickly.

A commercially available MMH/NTO system is considered a low thrust system, but by comparison to MPD's we had better call this a medium thrust system. A hydrazine based system, it has an Isp of 325 producing a thrust of 503 newtons, (the MPD = 31 N). As one possible model, at a cost of 1.8 tons of MMH/NTO engine mass and 14.3 tons of fuel we could add on the first 1000km of orbital altitude in one day and save 107 days off the one beaming station time line. A 23% saving off of whatever number of stations are employed. Seems like a good idea. The MMH/NTO booster system only adds 14.6 tons (its fuel) to our mass because its engine mass can be substituted for some of the extra counterweight mass we are sending. For multiple beaming stations this saves 35 days with three beam stations and 26 days with four. This would indicate that about the best we can do is 126 to 137 days to GEO, plus 71 more days to end of mission.

Notice that while we have been finding ways to speed up a slow starting mission we still leave unaffected that 71 days of post GEO maneuvers hanging over our heads. (Intended.) There is one other thing to consider that can lower the time line on both ends of the mission. We can design the MPD engine to run at a lower Isp, which is to say at a lower performance, using more fuel at a higher flow rate. As we pointed out at the beginning of the discussion of electric engines their marvelous efficiency comes from the slow fuel flow rate at high exhaust velocities, the trade off being, it takes a long time to get somewhere. (Remember that liquid rocket engines, Isp 450, get there in one day, but need a lot of fuel.) If we decide that we don't have quite that much time we can run the fuel at a higher rate, which increases the thrust but lowers the Isp, (the efficiency) and gets there sooner. (A lower Isp engine might also be chosen for design reasons. The 4000 Isp engine is based on lithium fuel, which can contaminate and damage surfaces. If this does turn out to be a problem, other fuels are readily available, nitrogen, neon or argon, all of which are inert but would produce a lower Isp.) This uses more total fuel and increases the total mass of the mission, which is the basic reason for not doing it.

The actual mass of the final components and the budget will decide. The question being, what will it cost to get the job done sooner, and what will we save in doing so? If a lower Isp is chosen it could be anywhere between 2000 and 3500/s. Here is what an Isp 2500/s mission profile might look like, compared to Isp 4000/s.

Table 5.5: MPD, with MMH booster, Mass Breakdown, Isp comparison

	Mass (tons) Isp 4000	Mass (tons) Isp 2500	Requirements on subsystem
Payload (ribbon)	40.0	40.0	
Spools and Structures	10.0	10.0	Support ribbon mass during launch
Spacecraft frame, command and control, panels, etc.	3.0	3.0	Mounting and control systems for flight and deployment
MMH/NTO system.	1.8	1.8	Move from LEO to 1300km
MPD Engine system	2.0	2.0	Move from 1300km to GEO
Fuel tanks	2.0	2.0	Tanks with pumps, both systems
Additional Counter Weight mass	10.6	10.6	Can be used for additions to systems above
Total Dry Mass	69.4	69.4	
MMH/NTO booster fuel.	14.6	**16.2**	Fuel for LEO to 1300km altitude
MPD fuel	13.0	**22.0**	Fuel for 1300km to GEO
TOTAL	97.0	107.6	

The post GEO maneuvers would be shortened from 71 days to 45 days and the trip to GEO would be shortened from 137 days with three stations to 91 days. Overall, for three beaming stations, 208 days reduced to 136, a saving of one-third and a long way from the 552 days at the start. Still, we should pause for a moment in consideration of the poor operations people that will have to be on duty for 136 straight days and nights!

There are many variations on this theme of trading off thrust for time of mission. You will notice that the masses in the Isp 2500/s column goes over what we can cram onto 4 Delta IV Heavies (92 tons). So we may need to use five launches or get some of the "additional counterweight" from the launcher itself. Or we could retreat all the way back to using a small high thrust, liquid fuel engine (for its lower, 450/s Isp) in place of the MMH booster to get the trip off to a fast start. This could save us up to 5 tons and/or get us out further than the 1300km orbit in the first jump. Many tradeoffs to be considered.

Chapter 6
Anchor

One of the major tradeoffs of the space elevator program and one of the more critical is selecting the location of the anchor. There are both political and technical aspects to selecting an anchor location. The technical selection considerations include:

1. Global distribution of lightning activity (see Subsection 10.1 Lightning)
2. Global distribution of cyclonic storm activity (see Subsection 10.4: Wind)
3. Global distribution of smaller storms (see Chapter 4: Power Beaming)
4. Requirements associated with locating a power beaming station near-by (see Chapter 4: Power Beaming)
5. Available real estate to allow for mobility of the ribbon anchor
6. Ease of construction, access and operations
7. On, or very near the equator, (at least for the first few decades).

When considering these criteria we come up with a short list of locations. The first possible anchor location is a floating platform located off the coast of Ecuador. The second possibility is a mountain-top location either in Ecuador or Tanzania. At this point we feel that the first option is far superior to the second.

Locating the space elevator ribbon anchor on a movable, ocean-going platform has numerous advantages over a land-based anchor. These include:

1. Excellent mobility for moving the ribbon out of the path of low-Earth-orbit objects and any storms strong enough to warrant taking evasive action.
2. Can be located near the equator in an area with very few lightning strikes, no cyclonic storms, few overcast days, and calm weather (~1500km west of Galapagos islands).
3. Can be located in international waters
4. Can be located near populations or not as selected
5. Large-scale, mobile sea platforms are tested technology (oil-drilling platforms and specifically *Sea Launch*)
6. In the event of a break in the ribbon, the lower 1000 – 2000km of the ribbon would land in the ocean, the ribbon segments above this would likely burn-up on re-entry.

7. No high-altitude operational challenges such as: snow, construction on or near glaciers, difficult access by land or air, limited usable land area, and breathing difficulties due to reduced oxygen.

8. Being on the sea is both an advantage and a disadvantage (see below). Shipping large or out sized objects is easier on the sea than the land. We easily build objects on land that can't be shipped, such as buildings, by shipping in stacks of the much smaller building materials. In space we would rather not build things on site, we would rather they arrived whole or in large pieces needing minimum final assembly. Being on the sea we can ship large and ungainly objects or their pieces that we could not transport to a land site. The larger the ribbons the more this is an advantage.

The disadvantages include:

1. Movement of the anchor. The sea is not as steady as the land, tides and ocean waves can move the largest ship when we may not want it moved, and we want to be able to move the ship and drag the ribbon to a different spot as needed. Unintended as well as intended movement could exceed the ribbons weight limit and cause a separation. The ribbon will have to be mounted in a motion compensating system to prevent ship movement, mostly vertical, from putting more tension on the ribbon than its limit. This system must be able to manage the ribbon tension during the starting of a climber, the most critical period of operations when tension margins are lowest.

2. Movement of the power beaming station. When power beaming is located on separate facilities, a similar motion compensating systems will be needed, this time to keep the beam steady on the climber. A wandering beam would cause power interruptions and a rather jerky ride.

3. Salt. On, or even near, the ocean, salt spray will occur. As any frequenter of salt water environments knows, salt can be very corrosive. Ships are generally built to deal with this, but most of our installed equipment will have to be built to "hardened" standards. Climbers and equipment that come in contact with the ribbon will have to be stored in air-conditioned or other suitably clean conditions. The first hundred meters of the ribbon may have to be washed down periodically.

4. Remoteness. The lack of normal homelike and civilian diversions may require additional facilities, as yet unplanned, as compensation.

Fig. 6.1: Odyssey from the Sea Launch program.

A good starting point for considering the various aspects of an ocean-based anchor is the *Sea Launch* program. The launch platform (*Odyssey*) for the *Sea Launch* program (figure 6.1) has the following characteristics [Emberly, 2000]:

1. Refurbished oil platform
2. Overall dimensions: 133m x 67m
3. 46,000 tons displacement
4. Power plant: 26,800 horsepower (20 MW)
5. Self- propelled transit speed: 12 knots
6. Cost: <$100M (our unofficial estimate)
7. 18 months for refurbishment
8. Home port: Long Beach California (already in the Pacific)

Based on the *Sea Launch* program a movable, ocean-going anchor platform could be fairly straightforward. The existing platform has:

1. sufficient mass to not be affected greatly by the 20-ton capacity ribbon
2. sufficient mobility to address our orbital collision avoidance concerns
3. sufficient power and real estate to accommodate a power beaming system
4. facilities for a substantial crew

5.   the flexibility to accommodate additional needs
6.   already been tested in the Pacific Ocean

In the future the anchor facilities can be expanded almost indefinitely with additional floating platforms (independent or possibly interconnected).

In general the oil industry is our guide in building large, they have built the largest movable objects in the world, oil tankers and ocean drilling platforms well in excess of 200,000 tons. When we get around to needing larger anchor ships they will have the equipment for us to use. For example the drilling ship Discovery Enterprise; 110,000 tons, 254 meters, with 42,000 horsepower in 360 degree directional engines for station keeping or cruise, and another 30,000 horsepower for the drill works. Prices upon request.

Chapter 7
Destinations

Space traffic on the ribbon can be put into three categories. LEO of several altitudes (some divide this into low and high LEO (LLEO and HLEO), GEO for "stationary" satellites, and past GEO for "Escape" trajectories that go someplace else. The ribbon can service all of these categories depending on the portion of the ribbon being used. The ribbon sections below GEO, (35,785km) services all of the LEO orbits, at GEO obviously services that category and the sections beyond GEO altitude can sling ships to other worlds needing only rocket engines and fuel for orbital corrections. The vast majority of this traffic, with today's rockets, is going to some orbit around Earth. First going to LEO (low) where they use a second rocket burn if they are going higher. So, the usual economics is that the Shuttle can deliver 25 tons to LEO, but if it is going to a higher orbit that 25 tons has to include the second rocket and its fuel. The result is that only about one half or as little as one quarter of that mass winds up as the payload in the higher orbits.

If you drop a load off of the ribbon, short of GEO, it will start an elliptical orbit with the drop point as the high point of the ellipse and the low point at or close to Earth. Climbing the ribbon does gain some rotational velocity, but not necessarily enough to stay in orbit once released. The point of departure from the ribbon must be high enough that the low point of the ellipse misses the Earth and the atmosphere. Alternatively, the load being sent to LEO has to include a rocket motor and fuel to, first, kick it up to the needed low orbital speed and second, to cut the ellipse down to a circular orbit.

Fortunately it takes much smaller motors and less fuel to insert a satellite into orbit from the ribbon than it does from the ground, somewhere around a 300 to 1 advantage. The higher up the ribbon we go the less extra energy we have to apply with rockets. At GEO the rotational velocity is 3.07 kilometers per second (km/s), which is the same as the orbital velocity for that altitude so the satellite stays there. Rotational speed drops below 3km/s as you move lower on the ribbon (from GEO) and average LEO speed is 7.6km/s so you can see that we have about 4.6km/s to add to our payload, once dropped. This has to be done in two stages. The first burn, as soon as we drop off the ribbon, is to raise the low point of the ellipse to the LEO altitude we want. (We are "falling" and we would rather fall further east and miss the Earth and the atmosphere entirely.) The second burn is done when the ship gets to that new low point, on the opposite side of the Earth, in order to circularize the orbit at that altitude; otherwise it would return, near to the high point of the ellipse it started from.

Our working payload on the 20 ton climber is 13 tons so some of this has to be for engines, tanks, and fuel to insert the satellite. Let's say we want a 300km LEO. From a 14,000km drop point the two burns take about 7 tons of fuel, 1 ton for engines and tanks leaving 5 tons for the final satellite. As we go higher, the economy gets better. At 20,000km it only takes 5.67 tons of fuel leaving 6.3 tons for the satellite, and at 23,750km it takes less than 5 tons of fuel leaving 7.1 tons for the satellite. At 23,750km an interesting thing happens, there is no need for the first burn if you want a 300km orbit, because left to fall by itself the low point of its ellipse is 300km. (If you want a different altitude orbit you would have to use first burn or another altitude on the ribbon.) In other words, from 23,500km on up we miss the Earth and atmosphere and go into a widely elliptical orbit.

We could put the whole 13 tons there with no need for engines and fuel but there isn't much call for satellites in such looping orbits. We could sell one-day rides "A fall around the Earth" that come back and reattach to the ribbon, that might be a thrill. The advantage of going higher on the ribbon for insertion drops is not only a larger payload left to orbit but the greater reliability of one burn and also ribbon altitude comes cheaper than extra rocketry.

Scalability

Another interesting concept that is different with the ribbon is scalability. Rocket launches from the ground do not scale; that is, you can't just use smaller rockets for smaller loads. It takes a good-sized rocket just to get anything into orbit then you work up from there. But the ribbon does scale, in the sense that you could run something just a short way up the ribbon and fire away, you don't have to go to some great altitude first. Climbing to just 500km our 13 ton working load could fire off horizontally, and put a 1.4 ton satellite in orbit using 10.4 tons of fuel. We're not suggesting that this is always economical but it is a choice that you don't have with rockets. This would definitely become an option if it developed that the company could charge by how high you go, because the climber could be returned sooner for another payload. (That may seem like the obvious way to charge for services but that's not the way it will work at the start. Prices will be set to pay off the start up costs of the ribbon, not just to run the taxi. See chapter 13 on the economics.) These days, as technology makes things smaller, lighter and yet with more features, you can do a lot with 4 or even 1.4 tons of satellite.

A major difference with using the Elevator for lifting satellites is that for the first time satellites will be coming <u>Down</u> into an orbit instead of going up to one. This will allow for a certain inventiveness, nay, even competition for alternative arrangements to achieving a circular LEO for various cost factors. As mentioned above, the first burn (where we speed up to establish the low point of the orbit – well above the atmosphere) can be eliminated by dropping from above 23,000km. This may well become standard practice, time on the ribbon for the extra altitude being cheaper than the rockets and fuel necessary for that first burn. This being the case, one way to eliminate the second burn, (at the low point of passage, to circularize the orbit) is to substitute an aero breaking maneuver. After first supplying the new satellite with a suitable aero breaking shield, (a "hat" to plow into the air) we make the drop so that we <u>don't</u> miss the atmosphere, we skim into the thin outer regions and slow down by friction, the same way many Mars probes are handled.

Another method, again starting above 23,000km, is to use a "tether drag" maneuver. A tether is a length of conducting ribbon (this ribbon could be of conventional materials used currently or, preferably, our new nanotube materials) that is deployed so as to drop below the satellite's position. This does two things; it interacts with the Earth's magnetic field, inducing drag and also a gravity tidal differential, that creates a drag on the satellite. This will <u>eventually</u> slow it down to the proper orbit, at which time the tether is discarded. The tradeoff is time. We have something on the order of 3km/s ΔV to dissipate at the low end of the ellipse and the tether is going to be supplying drag in the range of less than a meter per second. The amount of drag depends on the length of the tether, so we may need tethers tens, even hundreds, of kilometers long to get enough drag force to slow down the satellite in the customers lifetime. Someone can do the math later, these are just some of the different ways "spacecraftsmanship" will likely develop with the Elevator. In any case be sure to order your tethers from Tethers Unlimited, Seattle Washington, a subsidiary of ... well you get the idea.

To someplace else

An additional aspect of the space elevator is that the same system can be used as a sling to lift payloads to more distant destinations, whereas rockets require complex second and even third stage systems.

To find out which solar orbits are accessible by the rotational "throwing power" of the space elevator we have to get into the complexities of the conservation of momentum and energy and orbital mechanics. The basic idea is simple enough, take the difference in momentum between two orbits, such as Earth and Mars, and that is what you need to launch your trip. But the details get a bit sticky. For one, you have to add an extra factor for getting out of the gravitational influence of the Earth before your trip can be truly underway. And second, when using the elevator you have a variable starting condition. That is, each altitude on the ribbon gives a different gravity-well condition that you are escaping from. This makes the final equation complex, with two unknowns or mutually dependent variables, so that the only convenient way to solve it is to use a spreadsheet and iteratively drive the two variables to a common solution. (The math used for our current rocket trips is easier, we have a fixed starting point, either Earth's surface or a given LEO orbit, so this problem does not arise.)

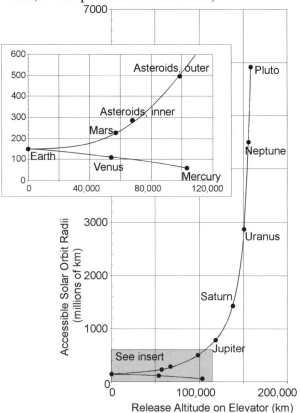

Fig. 7.1: Accessible solar orbits from the space elevator. The orbits of the planets are marked along with the ribbon required to access each.

For those that are interested in the details please see the section in the addendum on "Calculating Delta V's for interplanetary Orbit Insertions." Here we present figure 7.1 and table 7.1 displaying the visual results of the calculations. For an idea of how this works, a simplified version of the math would look like this:

$$R_r = V_r*(86154)/2\pi$$

That is, R_r (the needed altitude on the ribbon) = V_r (the velocity needed for the trip) * (times) (86154) (the seconds in a sidereal day) $/2\pi$ (divided by 2π). V_r is the velocity of the ribbon (at a specific altitude) and has to equal the "insertion velocity" for a given destination. For Mars it is 4.6147km/s which, from the formula gives a ribbon altitude of 63,276km, from the center of the Earth – so subtract the Earth's radius, 6378km, to get the more commonly used altitude of 56,898km from the ground, as shown in the table.

Table 7.1: Orbital distances and release altitude from the ribbon.

Planet	Average Distance/Sun Millions km	Minimum Altitude On Ribbon
Mercury:	*57.9*	*103,348*
Venus:	108	54,148
Earth:	149.6	0
Mars:	228	56,898
Asteroids, inner	285	67,748
Asteroids, outer	495	98,748
Jupiter:	778	119,063
Saturn:	1,430	138,418
Uranus:	2,870	151,383
Neptune:	4,500	156,322
Pluto:	5,910	158,441

Note, heavy line = free ride limit of first ribbon's length. Note that Mercury is not included.

With this formula you can get the ribbon altitude for any departure velocity you want. The problem, of course, is where do you get the velocity needed for a given destination such as the Mars figure we used. You can fudge it by adding about 25% to values from a table of delta V values for trips from LEO to the planets or do the math as in the addendum with an iterative spreadsheet.

Timing is important. To go to the outer planets, Mars etc. you cut loose as you come around the Earth with the Sun on your left, facing the direction of motion. The Earth's orbital speed around the Sun plus the Earth's rotational speed, now enhanced by the swing arm of the ribbon, give you the overall energy to go into a higher orbit around the Sun. (Mathematically, we also have to allow a little extra for the retarding effect of Earth's gravity as we start our trip; picky details.*) To go to the inner worlds, Venus and Mercury, you have to let go of the ribbon with the Sun on your right, facing direction of motion. We are subtracting the swing arm speed of the rotating ribbon from the Earth orbital speed, in order to go slower, relative to the Sun and drop into a lower orbit around the Sun. You will notice a curious thing in these numbers, even though the orbit of Venus is closer (42 million km) to Earth than Mars (78 millionkm), it takes about the same energy, that is, ribbon altitude, to go to Venus. (These are average orbits; in some conditions it would take more ribbon altitude to go to Venus than to Mars.) This is because the gravity well steepens as we go closer to the Sun, needing more energy for a small change in orbit as opposed to going away from Earth orbit, where a given amount of distance takes less energy.

While the ribbon does the major work in getting us on our way, it is not entirely a free ride to our destination. Keep in mind that the ribbon, being on the equator, is swinging around Earth in the equatorial plane, tilted some 23.5 degrees to the plane of the planets of the solar system, the ecliptic. These calculations are for minimum energy transfer orbits called Hohmann* orbits in the plane of the ecliptic. Therefore, since the ribbon lifts in the Earth's equatorial plane, we need to carry rockets and fuel to make a "plane change", a ΔV to bring the payload back into the plane of the solar system. Engines will be needed anyway for midcourse fine-tuning of the trajectory and to circularize the orbits at the destinations.

The amount of ΔV needed for the plane change maneuver varies with the season, from a maximum of 1.282km/s down to zero. The tangent to our equatorial plane, the circular plane of ribbon rotation, is the path a ship would take when released. Twice a year this tangent, at the outer edge of our circular plane, furthest from the Sun, points in line with the ecliptic. If our target happens to be in the right position to make that trip, at that time, we can forgo the plane change. These figures are for average conditions. Gravity assists have not been considered here, nor have the eccentricity of the Earth's orbit and other target destinations. For instance, the eccentricity of both Earth and Mars orbits bring them as close as 54.5 millionkm where we could launch from a 53,539km altitude on the ribbon or as far as 102.1 millionkm where we would have to go out to 60,068km on the ribbon.

*[For an explanation of Hohmann orbits and how to compute the insertion velocities that translate into altitudes on the ribbon, see Addendum.]

For our initial ribbon we will constrain our ambitions and select a ribbon that will allow access to Venus, Mars, and the asteroid belt. This ribbon length will be 100,000km. We can still launch to other destinations; such ships will just have to include fuel for the extra ΔV, not yet supplied by the ribbon. The Space Elevator would still be providing such a large starting advantage that this extra requirement is not that great an impediment to outer system exploration. Once the first elevator is established, longer elevators can be constructed so the outer planets and Mercury can be reached more easily.

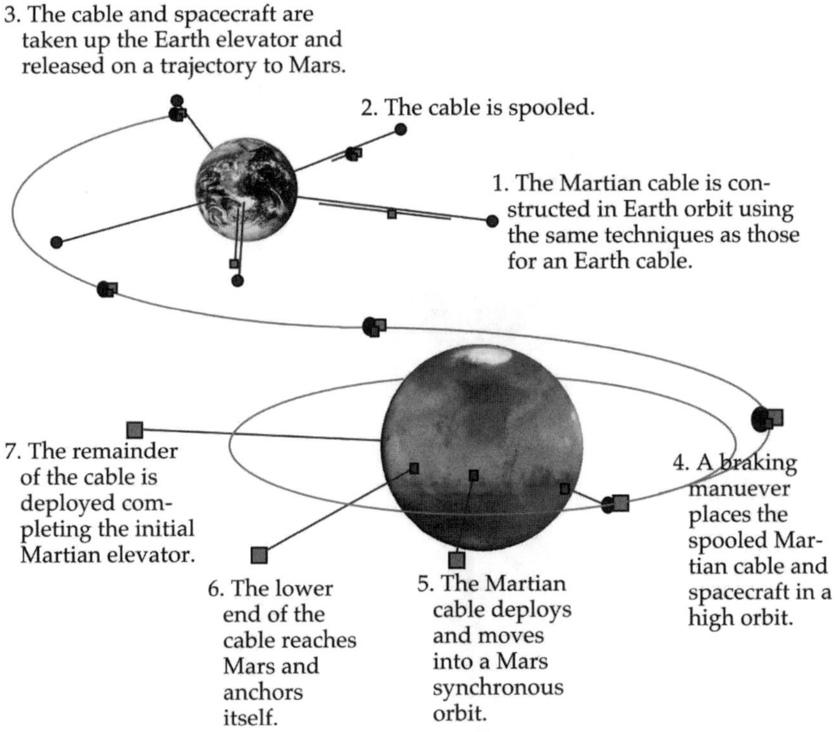

3. The cable and spacecraft are taken up the Earth elevator and released on a trajectory to Mars.

2. The cable is spooled.

1. The Martian cable is constructed in Earth orbit using the same techniques as those for an Earth cable.

7. The remainder of the cable is deployed completing the initial Martian elevator.

4. A braking manuever places the spooled Martian cable and spacecraft in a high orbit.

6. The lower end of the cable reaches Mars and anchors itself.

5. The Martian cable deploys and moves into a Mars synchronous orbit.

Fig. 7.2: One possible scenario for deploying a Martian elevator.

Martian elevator

One additional use of the space elevator is production and delivery of ribbons for use on other worlds; for instance a Mars elevator (figure 7.2). The Mars ribbon could be produced in Earth orbit alongside an Earth elevator then spooled up and released as a single unit on a trajectory to Mars. Upon reaching Mars a braking rocket or aerobraking would be required to place the ribbon in the proper Mars-synchronous orbit (17,040km). From Mars synchronous orbit the ribbon would be deployed and anchored. The counterweight or a second package sent to Mars would be a space-based power beaming station. This power beaming station could utilize large solar arrays, solar concentrators or nuclear power and a simpler rigid mirror, since we won't need the AO precision in the thin Martian atmosphere where we will be sending the power down through the atmosphere, which greatly reduces the distortions.

Once the Mars elevator is established transport from Earth to Mars and Mars to Earth can be done with only a plane-change correction rocket, attitude adjustment thrusters and climbers. For example, a climber can ascend the Mars elevator to its upper end where it releases at the proper time to acquire a trajectory to Earth. When the Mars craft approaches Earth, it attaches to the Earth elevator (the proper positioning would give an almost zero relative velocity between the ribbon and the craft) and descends the Earth ribbon to the ground.

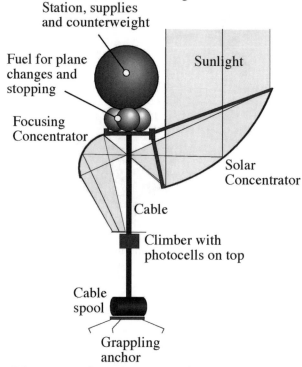

Fig. 7.3: Schematic of a Martian elevator showing the various components in one possible design.

Due to the Mars/Earth differences (lower gravity, lower synchronous orbit) the Martian ribbon would be roughly 1/2 the length and 1/3 the mass for the same capacity. The Mars elevator would have a different taper profile and not have to concern itself with lightning or man-made space debris. However, studies would have to be done to address possible Mars specific problems such as dust storms and the avoidance of Mars' moons. For a detailed discussion of the Mars ribbon, problems and possible solutions, see chapter 14.

Chapter 8
Safety Factor

The design of the ribbon for the space elevator straddles a fine line between impossibility and too fragile to survive. On one hand the ribbon can be designed to be strong enough to survive any problem and have orders of magnitude more strength than theoretically required. (This is, unfortunately, the trend in engineering practices today, a reflection of the zero risk attitudes. However, there is no zero risk; in engineering or in life. To demand it is a pretty good indication that some other motive is involved.) The problem is that such a ribbon is so massive that there is no feasible way to deploy it in a reasonable time, or at a budget that would get accepted in the first place. On the other hand, driven by expediency and cost savings, a ribbon can be designed skirting the theoretical lower limits in strength and end up coming down before the first climber begins its ascent. The problem is to find any middle ground that is feasible.

In the proposed system a standard safety factor of 2 was selected. This implies that the ribbon has twice the strength theoretically required at any point along its length. The question of a safety factor also changes as the ribbon is constructed. The ribbon is the most vulnerable at the very beginning but as it increases in size and thickness the stresses, from such factors as meteors and wind, diminish considerably.

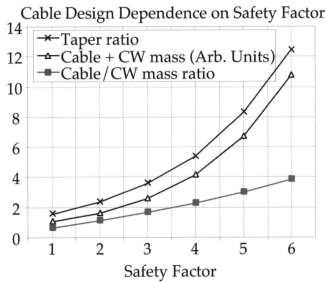

Fig. 8.1: The dependence of various ribbon attributes on the selected safety factor. (CW stands for counterweight)

The trade-off on the safety factor must be between the probability of catastrophic damage and what can be built at a cost that has the chance of a reasonable return on investment. In figure 8.1 we can see how the taper ratio and masses of the space elevator system depend on the safety factor. The mass has a direct impact on the schedule as well. Keeping the power beaming and climber attributes the same, a ribbon with a safety factor of 6 would take approximately 7 times as long to build as one with a safety factor of 2. In the proposed scenario this would be 17 years instead of 2.5 to put up the first 20-ton capacity ribbon. Improvements in the power beaming may reduce the absolute time schedule but the ratio is the same.

As can be seen in the diagram, the safety factor, an abstract concept, becomes concretely expressed in the ribbons taper ratio, how large at the top compared to the bottom. This in turn is directly reflected in the ribbons mass. The relationship between safety factor and ribbon mass is exponential, meaning that to get an additional increase in safety factor requires a large (disproportional) increase in mass.

In the chapter on challenges we discuss the most likely, and some unlikely, causes of damage to the ribbon. The actual taper ratio (the primary characteristic of our defined safety factor) does not come into the risk determination for any of the possible hazards. In other words increasing our taper ratio (or safety factor) alone is not the best method to reduce the risk of damage from meteors, atomic oxygen, wind, etc. Other modifications such as coating the ribbon or modifying the width and cross sectional dimensions or simply doubling the width of the entire ribbon will reduce the risk of catastrophic failure and these should be implemented where feasible.

From our understanding of the problem we find that increasing the safety factor (taper ratio) of our ribbon will greatly increase the construction time and costs without increasing the likelihood of our ribbon surviving. Other specific modifications reduce the risk of problems with much lower impact on construction. Note that, in any case, one would not build to a 4 or higher safety factor, it being cheaper and safer to build a pair of ribbons to carry that capacity while expending less mass of materials.

Chapter 9
Design Options

In the design, deployment and use of the space elevator there are various choices to be made and each of these has an impact. Below we have tried to show the impact of a range of possible options.

	Modification from baseline	Positive Impact	Negative Impact
1	Construction of Shuttle-C or other larger launcher	1. Larger initial ribbon and reduced risk of meteor damage	1. Shuttle-C development costs ($3B - $7B)
2	Use conventional chemical rockets to go from LEO to GEO	1. Off-the-shelf system	9. Higher cost 10. Smaller initial ribbon thus higher ribbon loss risk
3	Increase climber's power	1. Seven months less time to deploy ribbon 2. Faster transport to orbit	5. Thermal issues on climber 6. Motor mass hit
4	Increase the safety factor from 2 to 4	1. Once deployed the ribbon is less likely to be damaged by various factors.	1. 11 years to deploy first ribbon 2. Higher risk of meteor damage during extended initial deployment 3. Additional climbers required for deployment
5	Send up single 20 ton ribbon instead of a double 40 ton ribbon	1. Ease of deployment 2. Less development 3. Reduced launch costs	9. Added loss of ribbon risk 10. Additional deployment costs

In considering the various aspects of the space elevator program, the possible risks, the impact of an operational system, and its high public visibility, we recommend implementing design option 3 if viable but would suggest refraining from any of the other options that would increase the risk.

In addition to the design options stated here there are many other variations on the design that we have presented. These other variations are things like: a woven ribbon instead of one with vertical fibers and interconnects, laser beaming at 1.5 microns wavelength instead of 0.84

microns, 50 ton capacity ribbon instead of a 20 ton capacity, coating material to protect against atomic oxygen may be silicon dioxide instead of a metal, etc. These variations and trade-offs will come as the result of detailed engineering studies but are optimizations not really major design changes.

Chapter 10
Challenges

The primary influences on the space elevator design come from the environment that it must survive. The basic concept of a space elevator is straightforward but once you begin considering what will happen to it after it is deployed (and during deployment) you find that there are some design modifications that must be made. In this extended chapter we will touch on all of the major natural threats to the space elevator. The threats are in no specific order, some are extremely serious and some are of little concern but should be mentioned for completeness. With each threat we have attempted to supply a plausible solution.

Background Information
- Carbon nanotubes have a conductivity of 10^{-4} Ωm.

- The resistance of our ribbon (cross sectional area ~ 3mm^2) is 50kΩ or greater from cloud to ground (does not account for epoxy).

- Tethers have been used in thunderstorms and survived using Kevlar tethers. References to these programs have stated that the experiments were done when there was no rain. A wet tether or ribbon could change everything.

Subsection 10.1
Lightning

One possible event that would destroy the elevator ribbon would be a lightning strike. Lightning has sufficient current and voltage potential in its arc to heat and destroy any composite that we have been considering. Normal human structures such as tall buildings and especially our magnificent suspension bridges like the Golden Gate suffer lightning strikes all the time. Unless you are out on stormy nights, with a camera, lurking for spectacular photos, as some of us more intrepid souls have done, no one notices. This is of little concern because the total mass of steel making up the pathway to ground is more than enough to manage the sudden energy load without bother. On the other hand, wires and small cables that might be strung into the air for various construction reasons, including research to deliberately invite lighting strikes, often suffer catastrophic breakdown, for the smaller ones even vaporization. (Well, at least it is a quick and honorable death.)   In the case of our ribbon we have just such a relatively small structure, from less than one square millimeter cross-section at the start (smaller than household string) up to two square centimeters for the largest ribbon we are likely to build in the future.

One could argue that the carbon nanotubes (melting point ~6000°) would survive a lightning strike better than anything else we have yet strung into the sky and that there may be a similar high-temperature epoxy that could be used for the lower section of the ribbon. True, but that may not be good enough, we won't know for sure until we have a chunk of ribbon long enough to test. The one centimeter (cross section) and larger ribbons of the future may indeed be impervious, however, for our starting conditions, we consider allowing cavalier exposure to lightning a high-risk option and believe that there may be a better solution to the problem.

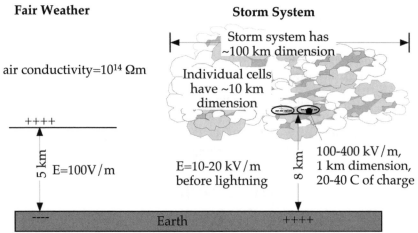

Figure 10.1.1: Illustration of the electrical properties of Earth's atmosphere.

The electrical properties of Earth's atmosphere are impressive, as can be seen on any stormy night or in figure 10.1.1. Potential differences of 400 kV/m can be produced with 40 C (coulomb) of charge stored in thunderstorm cells, (a potential of 16 million watts, now that would light up your life). The cells of potential and charge are isolated by the circulation of air masses so any part of a thundercloud could be charged even if another part is shorted to ground by current lightning. If we look at our ribbon, it could be a good conductor, at least better than air. This means the ribbon will appear to be the least resistance path to ground. In other words the ribbon will be the path lightning will prefer to take between cloud and ground, even to the point of "reaching" for the ribbon where lighting will jump a short distance through the air to the ribbon, rather than take a longer air distance to the ground.

The ribbon's conductivity could potentially be so good (better than copper) that it will drain potential charges before they can build to lightning size strength, if the ribbon is in contact during the build up. Unfortunately, the ribbon will not be able to sufficiently discharge an entire cloud having multiple cells. While the present cell in contact with the ribbon may be harmlessly drained of charge as fast as it can accumulate, other isolated cells with a full charge drifting into contact with the ribbon could discharge by lightning and damage the ribbon. If we decide to try to make the ribbon unattractive to lightning, with a higher resistance than air (a possibility) we might avoid a lightning strike by being the most resistive path to ground. However, rain often accompanies lightning storms and if the ribbon were to become wet the water may form a conductive path to ground. The lightning may take the water to ground and in the process enough of the lightning's energy may be imparted to the ribbon (through the EMF field or heat or explosive pressure or all of the above) to destroy the ribbon.

Possibly the best solution to the lightning problem is to avoid it, locate the ribbon anchor in a "lightning-free" zone like the one off Ecuador (figures 10.1.2 and 10.1.3). In such regions, lightning strikes will occur only a few times over the course of a year in a 100,000km^2 area. These strikes will also be concentrated in only several storms of rather limited size and extent. With statistics on our side the anchor station can easily manage to move the lower end of the ribbon out of the path of the few storms that do occur in these regions. This location and anchor movement scenario may also be required for avoiding high winds that may damage the ribbon.

A second alternative location for the ribbon anchor could be above 6 kilometers near the peak of a mountain. At least in one study it was found that there was a greatly reduced occurrence of lightning at these altitudes [Dissing, 1999]. This effect can also be seen in lightning frequency maps (figure 10.1.2). However, there are difficulties with locating the ribbon anchor on a mountain peak (Chapter 6: Anchor).

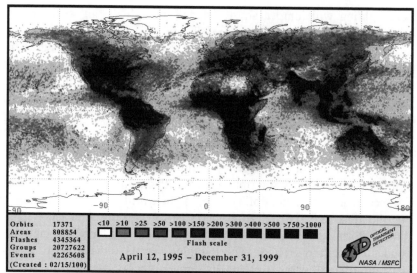

Orbits	17371
Areas	808854
Flashes	4345364
Groups	20727622
Events	42265608
(Created : 02/15/100)	

<10 >10 >25 >50 >100 >150 >200 >300 >400 >500 >750 >1000

Flash scale

April 12, 1995 – December 31, 1999

NASA / MSFC

Fig. 10.1.2: Global map of lightning strikes [Christian, 1999] (flashes/km²/year) showing regions of high activity (central land masses)and regions of low activity (eastern Pacific, northern Africa, and mountain ranges including the Andes and the Himalayas).

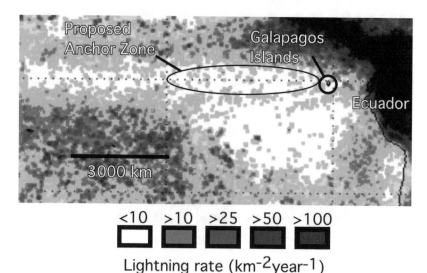

<10 >10 >25 >50 >100

Lightning rate (km⁻²year⁻¹)

Fig. 10.1.3: Expanded view of the "lightning-free" zone in the Pacific Ocean.

Subsection 10.2
Meteors

Meteors are a serious concern for all space hardware, and especially so for the survivability of the space elevator ribbon. Space is not as empty as we would sometimes like; it is quite busy, especially around large gravity wells like the Earth, with the small debris of rocks, pebbles, sand and dust. They come in very fast and from all directions. Fast, due to their own inherent velocity in their journeys through the solar system and also due to the Earth's considerable speed of 30 kilometers per second orbital velocity as we pass through the flight paths of these wanderers. (Average entry speeds are 12 to 36km/s, fastest possible, adding head on speed of Earth, ~72km/s. Faster than 42km/s and they would not still be in the solar system.) From all directions because they are the product of, and set on course by, the endless crashing and grinding of billions of meteors all over the system. They come in all sizes all the way down to dust only microns wide. The smaller they are the more numerous they are. The good news is that the large ones are easier to find and track, and there aren't so many of them. The bad news is that the vast majority are too small to find, much less track. But we can measure their behavior in the aggregate, as a group of objects, their size and frequency called their flux.

Meteor fluxes have been measured from Earth, and their impact characteristics have been studied in long duration space tests (LDEF) and in labs in high velocity impact facilities. It is interesting to note that even with the exotic technology of today's high velocity labs, using very long, high energy cannon, (far beyond the energy levels ever seen on a battle field) we can still only get test projectiles up to less than half the speed of the natural thing in space.

In the thick plate regime it was found that meteors could destroy a volume 50 times that of the impactor and to depths of several times the impactor's diameter, even in steel plate. Using only their mass, without the help of any high-explosive chemical, this is still better than our best armor piercing ordnance. The power comes from the speed; energy equals velocity squared, times mass. The fastest thing we are ever likely to have some familiarity with here on Earth is a rifle bullet, whose considerable force of impact comes from a speed usually less than 1km/sec. Meteors are steaming along at 10 to 50km/sec relative to the Earth, so when you square that velocity even a small mass can pack awesome force, 100 to 2500 times more than the rifle bullet. Much of the impact destruction is due to the energy shock that is created in the bulk material by the sudden conversion of all that speed, the kinetic energy to thermal energy.

In a thin plate scenario this changes. The shock to the immediate area of impact is actually more intense in a thin sheet solid because the reflection off the back face combines with the initial shock. However, much of the energy also escapes out the backside of the thin plate without the forces expanding to the surrounding area and destroying more of the plate. In effect, it punches holes instead of making large craters, so the total volume of material destroyed in a thin plate is less than in a thick plate. In our case we have a more unique situation, we have a sheet composed of independent fibers in a very thin plane. We will see how this affects our situation. But first we will examine our environment in more detail.

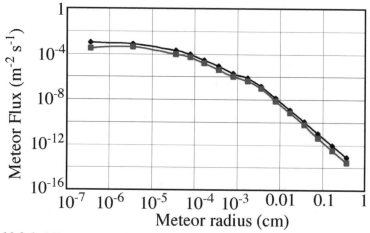

Fig. 10.2.1: Micrometeor data from Staubach, 1997. This data agrees well with that of Manning, 1959 which was used in the first, preliminary, space elevator paper [Edwards, 2000].

Published micrometeor fluxes from several sources give roughly the same distribution (see figure 10.2.1). Below about 1 cm radius natural micrometeors dominate the population of material near Earth, above about 1 cm radius man-made space debris is the major constituent.

From the published fluxes we can calculate the impact rate we would expect and the resulting damage. If we are to assume a micrometeor will survive long enough to go straight through the ribbon and destroy a section (worst case), we will have catastrophic damage for large meteors at any angle of impact and for small meteors at grazing angles across the ribbon face (figures 10.2.2 and 10.2.3). For objects larger than 1 cm diameter we see the impact rate on our initial, vulnerable ribbon is once in several decades (figure 10.2.4), so it behooves us to build out the initial ribbon to its full size as quickly as possible. Grazing impacts by small meteors nearly parallel to the ribbon's long axis will not cause catastrophic damage since they would tend to make vertical tears and not greatly diminish the cross section.

The criteria we find for which meteors will damage the ribbon can be expressed as:

$$r + \frac{r\sin(\phi)}{\tan(\theta)} \geq \frac{wf}{2}$$

where r is the radius of the meteor, w is the width of the ribbon, f is the cross sectional fraction of ribbon that must be destroyed to sever it, ϕ is the angle between the ribbon face and the incoming trajectory of the meteor (in the horizontal plane) and θ is the angle in the plane of the ribbon face between the meteor trajectory and the ribbon's long axis (in the vertical plane); together these two angles would describe any incoming trajectory (figures 10.2.2 and 10.2.3). Integrating over the relevant angles we can find the fraction of the meteors that can sever the ribbon as a function of meteor radius. Combining this with data on meteor fluxes (figure 10.2.1) we find how often we can expect the ribbon to be severed by a specific sized meteor (see figure 10.2.4). Examining figure 10.2.4 we find that the small, grazing-incident meteors could destroy our ribbon quickly. Initially, this does not sound good, but we need to understand the situation better to really determine how much of a problem we have. First, we assumed the meteor would pass straight through the ribbon destroying everything in its path. Second, we assumed the ribbon was perfectly flat.

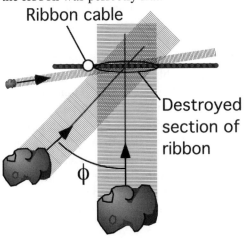

Ribbon cable

Destroyed section of ribbon

ϕ

Fig. 10.2.2: Interrelation between size, angle and impact damage. A small object coming in at a small angle across the face (left most object above) can damage as much width (not area) of ribbon as a much larger object coming straight on. The width of damage is more important than the area damaged because the intact cross section determines the strength of the ribbon.

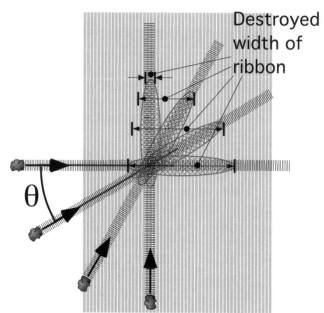

Fig. 10.2.3: Dependence of impact damage on angle of impact relative to long axis of ribbon. The destroyed width of ribbon is the critical parameter.

    When we look at our scenario, for example, we realize that some 100-micron particles are coming in at angles of less than 0.25 degrees to the ribbon face and continuing straight into the ribbon plane (one to ten microns thick on edge) for over two centimeters without being deflected. In many cases it will run into alternating regions of bare carbon nanotubes and epoxy/nanotube composite. This implies that the ribbon is flat to better than 100 microns across large sections of its face. What we have is a grazing impact on a thin sheet. It turns out that studies of this situation have been done for composite sheets [Lamontage, 1999: Taylor, 1999]. These experiments used impactors with roughly the same diameter as the target thickness and examined the affect of incident angle on the penetration, damage and ejecta. What was found was that the impact and debris were deflected out of the plane of the target and did not continue on their original path. It was also found that the damaged area and penetration depth dropped dramatically with increasing incident angles and with target thickness (the thinner the target the less area damaged). In other words the ribbon has some resistance to being punctured and will deflect some of the energy of small objects at shallow angles of impact.

    These experiments strongly argue that all impacts where the impactor and ribbon thickness are roughly of the same dimension (even grazing impacts) will damage no more ribbon area than several times the size of our impactor. This would place the puncture damage from this class of object in the range of one half a millimeter down to holes too small to see, a much more manageable situation than one might have expected at the start.

Now when we talk about impactors of 2 mm radius on our ribbon we are no longer in the same situation as in the published experiments because our impactor is now much larger than our ribbon is thick. In these cases further studies are required and since we don't have nanotube ribbon to shoot at yet we could run tests on thin sheets of Kevlar and graphite whiskers to get a class of results expected of nanotubes. But even from the present data it looks likely that the material will offer some resiliency and we will find that grazing incident impacts on our ribbon by meteors even several millimeters in radius will not stay in the plane of the ribbon and cause serious damage.

That being said, there is still some damage to be expected from this class of objects. Let's assume we still have a problem worth minimizing with 2 mm radius meteors (figure 10.2.4). One way to eliminate this hazard is to give the flat face of the ribbon a curvature, ripple or wave so that an edge-strike has no "plane" to cut across. With a non-flat structure we eliminate the problem of small, grazing incident meteors entirely, now they just punch holes, in the worse case a few holes millimeters in size. The wider the ribbon the faster the statistical risk decreases. That is one reason we are so interested in running two spools of ribbon on initial deployment. A key width is to get past 30 cm, where the risk drops off sharply, as soon as possible. Finished ribbons will be a meter or more wide and the threat from meteors becomes negligible.

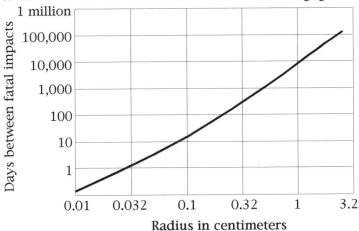

Fig. 10.2.4: Plot of the average number of days between fatal impacts on a 20 cm wide ribbon as a function of meteor radius.

Subsection 10.3
Low-Earth-Orbit Objects

Currently space debris larger than 10 cm diameter is tracked by U. S. Space Command. This accounts for roughly 8000 objects (satellites and space debris). An additional 100,000 objects with diameters between 1 and 10 cm are in Earth orbit (these are a separate class from meteors which are not in Earth orbit, they just zip on by). Of these objects most are in LEO (500 – 1700km) which has the highest and most deadly relative velocity to the space elevator ribbon. The difference in class of objects is important. Once a meteor has missed the ribbon it's gone, no need to worry about it again, but anything that is in orbit will eventually cross a vertical line through the equator, that is our ribbon; and then later come around and do it again. With this density of debris we can expect the ribbon to be hit and possibly severed once every 250 days. One possible solution to this problem is to duck, move the ribbon. This will require knowing when to duck, to track all of the space debris larger than 1 cm diameter and move the ribbon out of the path of any that are on a collision course (Chapter 6:Anchor).

Haystack observatory, a high resolution radar instillation, is beginning to study and track objects in Earth orbit down to 1 cm. Optical tracking systems are also coming on-line at this time. Tracking space debris down to 1 cm has been a concern of NASA because of its affect on the space station. A study was done at Johnson Space Center [Loftus, 1993] on the construction of a new debris tracking network and came up with a design that would monitor objects down to 1 cm with 100m accuracy using essentially current technology. This is very close to the tracking network we would need for the space elevator. An alternative system could be a set of five facilities located on the equator based on Berkeley's One Hectare radar Telescope. This would be an easily implemented and inexpensive solution.

Initially, while we are in the building phase, we want to avoid all impacts on the ribbon from objects larger than 1 cm (this becomes less stringent as the ribbon grows). Based on the system proposed by Johnson Space Center the space elevator would need to avoid a piece of space debris every fourteen hours on average (see table 10.3.1). With an understanding of ribbon dynamics, a good computer system and the proposed anchor facility (Chapter 6: Anchor) this level of active avoidance is feasible.

One additional design modification that could be implemented is to widen the initial ribbon slightly at orbits where the debris is highest (figure 10.3.1). The vast majority of the critical debris is located between 500 and 1700km altitude (Interagency Report on Space Debris). If we were to design the ribbon to be twice as wide for these 1200km we would reduce the risk of serious damage by roughly 30%. This pushes the critical size of meteor to be avoided up to roughly 3 cm, which is more easily tracked, with greater accuracy, reducing the frequency of avoidance moves. The increase in the ribbon mass would be 0.65% including the upper ribbon to support this extra mass. It would be expected that additional facilities and advancements in the sophistication of this technology would result in greater tracking accuracy and a considerable increase in the time between required ribbon moves and size of the move. In the end the 10 to 20 meters moved every 5 or 6 days would be not be noticeable, in fact those movements would be smaller than the station keeping movements that the ship would be doing anyway to offset the effects of wind, wave and currents.

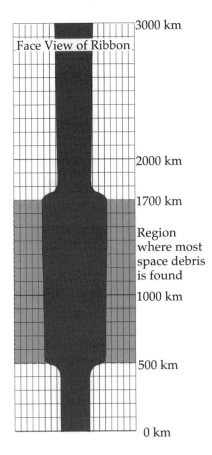

Fig. 10.3.1: Sketch of how the width of the ribbon may be modified to improve its survivability in regions where space debris is most prevalent.

Table 10.3.1: LEO Collision Avoidance of All Objects Larger than 1 cm

Tracking accuracy (m)	Time between required ribbon movement	Minimum size of movement required (m)
1000	1.4 hours	1000
100	14 hours	100
10	5.8 days	10
1	5.8 days	10

Subsection 10.4
Wind Loading Study

Ah, Wind, beloved of song, poetry and prose, the stirring of the atmosphere without which we would have no life-giving rain or life-allowing moderate temperatures. Beloved, except by engineers who have to erect structures on this Earth. Today's buildings and bridges abide the wind, with a few dramatic exceptions like the Tacoma Narrows Bridge, which is a useful reminder that there is only one kind of man-made structure that the wind has never blown down, the great Pyramids. (And some day in the far future, in a fit of frustration, the wind may just bury them in sand anyway.) For the most part we now know how to design for the forces of the wind, wind loading tables are available for everything, except things that have never been built before, like our ribbon.

Let's assume we have the initial and weakest ribbon deployed and a wind blowing across a 1km segment of its length. For a first example, we will also assume it is acceptable for the ribbon to be displaced, moved out of the vertical by the wind, by as much as a 10° deviation from its nominal position. The question is, what wind velocity will break our ribbon in this scenario.

The force from the wind perpendicular to the ribbon face required to break the initial and weakest ribbon is:

$$F = T \sin \theta = 900 kg \cdot 9.8 \frac{m}{s^2} \cdot \sin(10°) = 1531 N$$

To do the calculation correctly we need to calculate the aerodynamic drag on the two distinct material areas, the ribbon or set of strings or rods (the individual fibers in the ribbon) and the regularly spaced plates connecting the rods. In addition, the ribbon will rotate in the wind to some extent. However, we will start with a slightly simpler and worse case where the ribbon is face on to the wind. In this case, all of the fibers and composite plates see the full force of the wind. In reality there would probably be some shadowing of the wind as the ribbon turns in the wind as well as some turbulence and fluttering of the ribbon.

The drag of an object can be expressed as:

$$D = \frac{1}{2} C_d \rho A v^2$$

where D is the drag, ρ is the air density, A is the frontal area, v is the air velocity and C_d is the drag coefficient (1.28 for a flat plate, ~0.07 for a cylinder in low velocities)

The ribbon has two types of area, cylinders of fiber and flat plate composite sections, so you do the equation for each type, add them together, and then rearrange to solve for v. (For the rest of the math, see Note 2 at the end of this section.)

With the force required to break the ribbon from the first equation 1684, and 1200 fibers of 10 micron diameter (one possible configuration for the first, smallest and most vulnerable ribbon), with 5% of the length in "plates" and with the wind effective over a 1km vertical extent we get:

$$v = 32\frac{m}{s} = 116\frac{km}{hr}$$

When we do the math for the two types of material areas of the ribbon, it is easily seen that almost the entire drag comes from the composite sections even though it only represents 5% of the ribbon area. The reason for this is they fill the area between the fibers making for a much larger effective area per unit length and the drag coefficient for a flat plate is almost twenty times that of the cylinder.

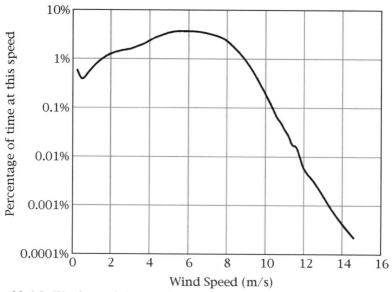

Fig. 10.4.1: Wind speed distribution at the proposed anchor location..

Looking at the historical record of wind speeds at the proposed anchor location [PACS data, Weller, 2001], (figure 10.4.1) we find the velocity distribution is actually considerably below the 32 m/s breaking velocity. However, it is always better to have a larger margin of safety, if we can easily get it. A second paper [Sandwell, 1984] gives a global map of the seasonal average wind speed. In the maps one can see the spatial distribution of high and low wind regions. The proposed anchor location

(~2000 kilometers west of Ecuador) is found to have low wind speeds year round and is also in our preferred area of very few lightning strikes. Things are coming together nicely.

Design factors that we can apply to further reduce the risk of damage by wind include:

2. Reducing the area of the composite section, that is, the ratio of nanotube/epoxy composite to bare nanotube length in our ribbon design. Since the composite sections will be creating the most drag, reducing the fraction of the ribbon they represent will reduce the total drag. Achieving this is dependent on the composite technology as yet to be developed.

3. Reducing the width to thickness ratio of the ribbon. (figure 10.4.2) By making the ribbon one fifth the width we reduce the wind load on the ribbon by a comparable amount and increase the critical wind velocity by the sqrt(5) or from 32 m/s to 71.5 m/s (159 mph or a category 5 hurricane, table 10.4.1).

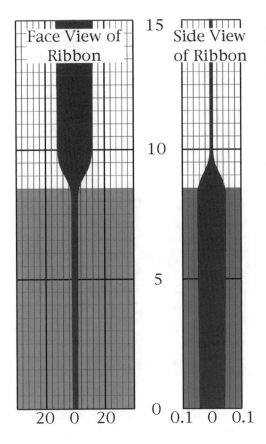

Fig. 10.4.2: Modified ribbon design to deal with wind loading. Vertical scale is altitude in kilometers, horizontal scales are in centimeters. The face wide decreases from ~20 cm to ~4 cm and the width goes from a single fiber layer to multiple layers.

Table 10.4.1: Storm Characteristics

Type of Storm	Category	Winds (mph)
Depression	TD	< 39
Tropical Storm	TS	39-73
Hurricane	1	74-95
Hurricane	2	96-110
Hurricane	3	111-130
Hurricane	4	131-155
Hurricane	5	>155

Location-Related Wind Considerations

Reducing the width to thickness ratio of the ribbon to the more stocky profile for wind resistance need only be done to the first 7 to 8km of ribbon that is in the wind and weather portion of the atmosphere.

If the design modifications push the critical wind velocity that the ribbon can stand to roughly 159 mph, as suggested above, then we are discussing destruction that could only come from a cyclonic storm. If we are looking at an ocean platform anchor then the storms we are talking about would be, specifically, category 5 hurricanes. Not only do we not want our ribbon and its anchor ship in a cat 5 hurricane, we for sure are not going to stay there to watch it prove that it can survive a cat 5 hurricane. The answer then is to go where the hurricanes are not, by considering the anchor location in terms of the spatial distribution of hurricanes.

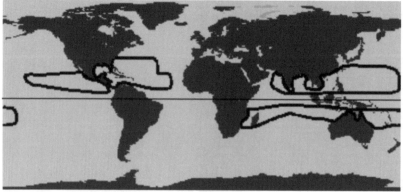

Fig. 10.4.3: Outlined regions indicate the location of hurricane activity.

Figure 10.4.3 shows the general and historical spatial locations of hurricanes. As can be seen in the global maps, hurricanes tend to exist at low latitudes, where there is plenty of warm water, in both the northern and southern hemispheres but do not occur at, or cross, the equator. This is because there are no coriolis wind forces on the equator with which to stir up, or sustain, a hurricane. It can also be seen that the eastern pacific off the coast of Ecuador has essentially no hurricane activity. This area is our current first choice for an anchor location based on the spatial distribution of lightning and may now solve our wind loading problems as well. This place has got to have the dullest weather in the world, Hurray!

Note 2:

We started with the expression for drag.

$$D = \frac{1}{2} C_d \rho A v^2$$

For a large number of individual fibers we then have:

$$D = N \frac{1}{2} C_{cylinder} \rho A_{fiber} v^2$$

where N is the number of fibers. For a flat plate like the interconnect sections we have:

$$D = \frac{1}{2} C_{plate} \rho A_{plate} v^2$$

The total drag will be:

$$D = \frac{1}{2} \rho \left(N C_{cylinder} A_{fiber} + C_{plate} A_{plate} \right) v^2$$

This can be solved for v:

$$v = \sqrt{\frac{2D}{\rho \left(N C_{cylinder} A_{fiber} + C_{plate} A_{plate} \right)}}$$

Subsection 10.5
Atomic Oxygen

Atomic oxygen is what one might call hyperactive! Divorced from their preferred mate, another O atom, (as in O_2) by upper atmosphere radiation, the single O atoms are in a permanent bad mood, dancing around looking for something to attack.

Atomic oxygen resides in the upper atmosphere between about 60 and 800km with the highest density near 100km altitude. It is extremely corrosive and will etch the epoxy in our ribbon and possibly the carbon nanotubes. On NASA's Long Duration Exposure Facility (LDEF) mission atomic oxygen etched carbon fiber/epoxy composites at rates up to 1 µm/month, preferentially etching the epoxy in some cases. This high etch rate was only seen on the leading face of the spacecraft where the atomic oxygen is being swept up by the considerable orbit velocity of satellites. On the trailing edge and in the shadowed regions the etching by atomic oxygen was found to be zero in many cases. The reason for this is that the oxygen atoms are not in orbit, they are just floating around as part of the wispy extension of the very thin upper atmosphere. The divorce from their partners under radiation bombardment also gives them a high thermal velocity of about 1km/s whereas a satellite's velocity is over 7km/s. The high relative velocity of the satellites driving through the atomic oxygen does for them what the front grill of your car does for bugs on a summer highway.

For our stationary ribbon we would not be adding to atomic oxygen's corrosive effectiveness by driving through it, we only have to deal with its own energy levels. We would expect that part of the etch rate would be down by about two orders of magnitude from what LDEF experienced on its leading edge. On the flip side, the LDEF results were measured at about 400km which places it at an altitude with an atomic oxygen density two times less than the maximum our ribbon will experience as it passes through the lower altitudes. The bottom line is we should expect to see etching rates of 1 µm/month in the highest density sections of the ribbon. This etch rate would be sufficient to destroy our ribbon in a few weeks in the current design (see figure 10.5.1). There are two possible solutions to this particular problem.

The first and probably best solution is to coat the affected segment of ribbon with a material that is resistant to atomic oxygen as suggested by many of the LDEF experiments. In addition to carbon/epoxy composites, LDEF also had bulk and thin film metal experiments, and metal-coated composites. During the 5.8 year life of the LDEF mission, gold and platinum were unaffected by atomic oxygen, while aluminum and several other metals were found to have minimal degradation. In the metal-coated composite tests it was found that coatings (nickel plus SiO_2) as thin as 0.16 microns could protect the composite from the affects of atomic oxygen.

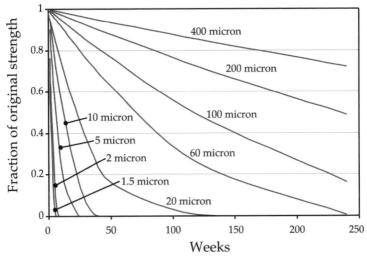

Fig. 10.5.1:Degradation by atomic oxygen of different diameter carbon/epoxy composite fibers as a function of time.

If we coat the affected length of the ribbon with a metal such as gold or aluminum we will need to make a trade-off between durability of the metal coating under the passage of climbers and minimizing mass so as to not weigh down the ribbon. A coating thickness between 0.02 and 25 microns may be acceptable depending on durability and the density of the coating. What needs to be completed are a set of tests that determine: 1) if the carbon nanotubes as well as the epoxy will be etched, 2) will a metal layer adhere to the epoxy and nanotubes of our ribbon, and 3) what is the minimum layer of metal that will survive the passage of several hundred climbers.

The second solution is to modify the ribbon geometry. The baseline, pre-modification ribbon has an average thickness of 1.5 microns and a width varying from 7.5 to 19.5 centimeters. This ribbon can be a uniform 1 micron sheet or made up of separate spaced fibers of 5, 10, 20, 30, up to 400 micron diameter. The larger diameter round fibers would survive much longer in the atomic oxygen environment. As can be seen in figure 10.5.1 unprotected fibers less than about 100 microns would be insufficient to survive in the atomic oxygen environment long enough for the ribbon to be strengthened. The tradeoff then becomes, these large diameter fibers may cause problems for the climbers, be more difficult to add fibers to and have a higher risk of damage by meteors (see 10.3, the round versus flat sheet damage propagation).

We have identified several variables here and mixed combinations of them leave more than a dozen ways this situation might be handled; we really need a lot of ribbon samples to test. Fortunately, the best approach, once found will only need to be applied to a relatively short section of ribbon.

Subsection 10.6
Electromagnetic Fields

The Earth has an active magnetic life. The liquid metal core, mostly iron, some nickel, churns and flows setting up the magnetic fields we know as compass headings. Then the Earth spins on its axis rotating these magnetic lines of force through space, setting up a shell of moving field lines around the Earth. The Sun does something similar, on a far vaster scale, of course, setting up the Interplanetary Magnetic Field (IMF). These interact with the Earth magnetic fields, making loops or lobes of magnetic fields around the Earth with the strongest densities orientated toward the Sun on the day side and an even larger lobe pointing away from the Sun on the night side. To liven things up a bit the Sun also puts out a rather energetic Solar Wind, a flow of ionized gas streaming outward from the Sun through all the magnetic fields. All together we wind up with a magnetic layer cake, stacked zones, spheres of plasma sheets, ring currents and radiation belts containing considerable energy surrounding the Earth.

On top of all this the Sun occasionally throws a hissy fit in the form of large solar flares, or coronal mass ejections, that send billions of tons of charged particles into space at speeds in the range of 1,700km/s, 91 hours traveling time Sun to Earth. These eruptions, riding with, and causing shock waves in, the solar wind impact the Earth's magnetic fields and for the most part are diverted around the Earth. If it weren't for the Earth's magnetic fields we would not be here, these solar eruptions are powerful enough to kill most organisms directly exposed to them. The reaction with the Earth's fields and the little that does get through do have effects here. They can effect our weather, knock out radio, TV and satellites, effect massive displays of northern lights and occasionally such strong surges in the electrical utility grids that they cause power outages over wide areas; an effect our ancestors, without electricity would never have noticed. Our space equipment and manned operations are built to deal with all but the strongest of these energy fields.

One might think that with all this electromotive force sloshing about that some future high technology could tap into it as a power source. After all, with the space elevator, we are waving a very long conductive wand around in large magnetic fields. Indeed, several science fiction stories have suggested just that, even to power one version of a space elevator. Nice try; as it turns out these energies are only dramatic and powerful on a world size basis. On the scale of the area that the ribbon affects, the forces are quite diffuse and they are quite variable, ranging over three orders of magnitude; not something you could depend on. But still, could these fields damage the ribbon in some way?

Other than the radiation danger that would exist even without the ribbon (see next section) the only direct effect is heat. Heating of the ribbon can be produced by passage through the local magnetic fields. The potential induced along the ribbon can be expressed as:

$$E = B(r)v(r)$$

where E is in volts/meter, $B(r)$ is the magnetic field, and $v(r)$ is the velocity of the ribbon relative to the magnetic field. For radii (r) less than 10 earth radii ($10r_E$), $B(r)$ is a very small number, and $v(r)$ is approximately zero.   However, if we assume the worst possible case where the magnetic field is fixed and the ribbon is rotating with the Earth we get potentials from 0.00026 V/m at $10r_E$ to 0.016 V/m at Earth's surface.   At distances of greater than $10r_E$ the ribbon is in the interplanetary magnetic field during the day   and is in the Earth's magnetosphere at night. (B_{ave} ~6 nT and B_{max} ~80 nT – very small compared with the magnetic field at Earth's surface of approximately 100,000 nT.) This corresponds to a maximum potential of 0.00068 V/m at the far end of the ribbon. With a minimum resistance of 0.4 Ω/m. we have a maximum of 0.0064 W/m of heating occurring near the Earth end of the ribbon and 1 μW at the far end; not enough to feel with your finger. The ribbon would quickly radiate this level of heating away into space; that is, the ribbon easily dissipates any induced heating much faster than it can be created.

The nut of it is that the energy fields are way too diffuse and our ribbon has very little velocity relative to these fields to have any noticeable electromagnetic effects.

Subsection 10.7
Radiation Damage

As with the electromotive forces, this is not a problem to the ribbon itself.

The segment of the ribbon in Earth's radiation belts will experience less than 3 Mrad per year (energetic electrons and protons) [Daly, 1996]. Studies of epoxy/carbon fiber composites (epoxy/nanotube composites would be expected to be comparable or better) have found them to be radiation hard to greater than 10^4 Mrad [Egusa, 1990: Bouquet, 1979]. This would allow them to survive more than 1000 years in the expected environment. In other words, we will wear them out long before we rad them out. The other components of the ribbon, composite matrix material and interconnects, will need to be selected to survive this radiation as well.

The other radiation damage that must be considered is that of solar UV. Often specific materials can be corroded by ultraviolet radiation, this won't effect the carbon nanotubes but this must be considered when selecting the matrix and interconnect materials for the ribbon construction.

Electronic, electrical, equipment and people using the ribbon would have to be safeguarded in the same way we do for using space now.

Subsection 10.8
Induced Oscillations

If you pound on something long enough and hard enough it will break. Other stresses such as tension and bending also break down objects. These are the basics of how oscillations can do damage. Oscillations, also known as vibrations, set up a variety of stress forces, as just mentioned, internal to the material that can first weaken the molecular bonds of the material and eventually break them. Even things that are meant to vibrate such as the strings of musical instruments eventually breakdown under vibrations they were designed for. Vibrations over 10 cycles per second, (hertz) are called sound. But lots of things vibrate at less than 10 hertz, and these cases are usually referred to as oscillations. Buildings and bridges are the usual concerns of oscillation studies, we have already mentioned the Tacoma Narrows Bridge in the wind loading section (10.4) as a prime modern example of failure in this regard. It wasn't the direct wind pressure, as in blowing a tree down, that destroyed the bridge, it was the built up over-stress of a pumped oscillation, induced by the wind, that did it. (By the way, don't blame the engineers too harshly, no one knew, before then, that oscillations of that size could be induced by the wind.)

A pumped oscillation is one where a periodic force is recurring in time (in step) with the natural bending or stretching moment of a structure, also called its characteristic frequency. We know about characteristic frequencies in daily life as the sound an object makes when struck. The tympani of clashing pots and pans, the ping of a water glass or a dropped spoon, the sing of tires on pavement and all the notes of musical instruments are the characteristic frequency of that object at that time. The pumped part comes from repeatedly applying a force to make the oscillation stronger, or larger. We recognize this in daily life by examples such as blowing or strumming a musical instrument harder to make a louder sound. A great example of pumping to the breaking point was the Memorex commercial of a few years ago where the glass was shattered by a loud musical note, pumping the vibrations of the glass beyond its tensile strength. "Was it live or was it Memorex?" The glass didn't care.

Buildings, especially tall buildings have characteristic frequencies and earthquakes are the perfect device for finding them. Most of the destruction in earthquakes is due to pumped oscillations, and most of the new construction techniques developed in California in the last 30 years are designed to allow motion without the build up of oscillations. The characteristic of the pumped oscillations was recognized over 2000 years ago, (without the name for it) when the Roman army instituted the rule that marching armies were to break cadence (walk in random step like civilians) when walking over bridges. Having built many bridges they found out that the in-step pounding of hundreds of feet could break down a bridge. (This rule only applied to their bridges.)

Our ribbon would certainly be the longest string in the world; would this make Earth the largest stringed instrument in the Galaxy? Will it vibrate? Most likely. What would be its "note", its characteristic frequency, and could forces pump the ribbon to the point of destructive oscillations?

Initial work by Pearson, 1975, on oscillations induced by the moon, sun and motion of climbers found the problems avoidable. However, since we have a fairly different system scenario the calculations need to be repeated. In Pearson's work, he had a ribbon 144,000km long with no counterweight on the end and he examined only taper ratios above three. Two factors will increase our characteristic frequency and one will decrease it. Our shorter ribbon will increase the characteristic frequency, the counterweight will double the characteristic frequency, from the case Pearson calculated, (by essentially fixing the upper end of the ribbon) and our smaller taper ratio will decrease the frequency. What is the net result of these changes?

If we examine the standard oscillation of a string under tension [Nagle, 1996] with the ends fixed, we find our system is very close to the ideal case. This assumes: no gravity, the string is perfectly flexible, the string is a constant linear density, the tension is constant and no other forces are acting on the string. In other words a guitar string, an understandable object. From these criteria it doesn't look like our ribbon qualifies as a true "string", but when you break down the physical elements, it actually does fit. (For instance the disqualifier, gravity, which we certainly have, disappears into the tension function. See math analyses in Note 3 at end of section.) If we have a true string then we can figure the characteristic frequency of our ribbon and move directly to see if any outside forces would disrupt it.

The solution to this problem is given in many differential equations textbooks and gives a characteristic frequency for our ribbon of

$$\tau = \frac{2 \cdot L}{n \cdot \alpha}$$

where L is 1×10^8 m, α is 7.1×10^3 m/s and n is 1, 2, 3, 4, ... The first mode has a period of 7.8 hours; this is the ribbon's characteristic frequency. This period is sufficiently far from 12 hours (the sun), 12.5

(the moon) or integral fractions of these so our ribbon should not be pumped by either the sun or moon.   Ribbons with lengths closer to 80,000km (6.25 hr natural frequency) and 160,000km (12.5 hr natural frequency) would be pumped by the sun and moon and could have serious problems.   Small oscillations or traveling waves that may be induced by wind or meteors can also be actively damped out at the base of the ribbon if a ribbon displacement monitoring system is implemented to detect any movements in the ribbon.

One oscillation that Pearson investigated was that of transverse waves induced by climbers.   The bottom line on this oscillation is that large oscillations can be induced when the climber transverses the length of the ribbon in one period of the ribbon's characteristic frequency. (Pearson assumed no counterweight so had the climber traveling twice the length of the ribbon during one period.)   Since we just calculated our ribbon's characteristic period to be 7.8 hours we will only need to worry about this particular affect when we plan to have climbers traveling faster than 10,000km/hr. The next century perhaps?

Note 3

The initial-boundary value problem for the standard problem is:

$$\frac{\partial^2 u}{\partial t^2} = \alpha^2 \frac{\partial^2 u}{\partial x^2}, \qquad\qquad 0 < x < L, \qquad\qquad t > 0,$$

(essentially F=ma for each segment of the string)

$$u(0,t) = u(L,t) = 0, \qquad\qquad t \geq 0,$$

(boundary values for the string having fixed ends)

$$u(x,0) = f(x), \qquad\qquad 0 < x < L,$$

(defining the initial position of each segment of the string)

$$\frac{\partial u}{\partial t}(x,0) = g(x), \qquad\qquad 0 < x < L$$

(defining the initial velocity of each segment of the string)

where u is our equation of motion, L is the length of the string, $f(x)$ is the initial location of the string, $g(x)$ is the initial velocity of the string (both of these before displacement by oscillations) and α^2 is equal to the ratio of the tension to the linear density of the string. This assumes: no gravity, the string is perfectly flexible, the string is a constant linear density, the tension is constant and no other forces are acting on the string.   This doesn't sound like our situation at all. Well, let's look at it a little closer. We have gravity, but that is what is giving us the tension and it is along the line of the ribbon in our case. For any individual segment of the ribbon, the force applied by gravity is extremely small compared to the tension, it is of the same vector as the tension, so mathematically it disappears into the tension. Our ribbon is pretty much perfectly flexible, the width is much less than the length, like guitar strings. Our ribbon does not have a constant linear density or constant tension. But let's look at where these two come into our calculations. Equation (1) is equivalent to F=ma for each segment of the ribbon. The only place where the design of the ribbon comes in is in α^2. And α^2 is also where the restriction on constant tension and linear density comes into play. The one unique thing about our ribbon - it is designed specifically such that the tension for any segment is exactly proportional to its cross-sectional area (or linear density). To be precise, to correct these equations we would need to put in a function of x in equation (1). However, because of our specific relation between tension and linear density this function is defined as 1 at all locations and has no affect on our problem. The initial boundary value problem we have stated above is a very close match to our situation. The last constraint was that there were no other forces acting on the string. To a large extent this is true, only the moon, sun and climbers will act on our ribbon. These are small forces that can pump oscillations but not dramatically change the characteristic frequency of the ribbon. [Wind forces act on the lower one hundredth of one percent of the ribbons length, that is not an aggregate force effecting the ribbon in the sense we are talking about here.]

Therefor: we can compute the characteristic frequency in the normal manner and use that to examine any outside forces acting on the ribbon.

Subsection 10.9
Environmental Impact

When considering the construction of a space elevator the possible environmental impacts must be examined. Two of those environmental impacts will be considered here. The first is the possibility of discharging the ionosphere and the second is the impact if a space elevator were to be severed and fell back to Earth.

Discharging the Ionosphere

First, what is the Ionosphere? It's the sphere where ions live.

In the upper atmosphere, ultraviolet and x-ray radiation strike the sparse air molecules, now mostly oxygen, knocking off electrons, thus ionizing them. The atoms become ions and the electrons become free roamers, in a charged environment. This is good, most of the ultraviolet and x-ray radiation is prevented from reaching the Earth's surface and harming us. Also, radio signals can bounce off the under side of this layer, propagating the signal around the curvature of the planet for long distance transmissions. The down side is that this layer also reflects away a large portion of the radio waves the radio astronomers would like to be able to receive from deep space.

The three major regions of the Ionosphere are: the D region at altitudes from 50 to 90km, the E region from 90 to 150km and the F region from 150 to 400km. (Whatever happened to the A, B, C regions, we have no idea.) The Sun is responsible for this process so at night the electron concentration drops, especially in the middle, E region, because without the incoming flow of energy things calm down and the electrons have a chance to recombine with the ions, becoming uncharged, respectable atoms once again. The top layer, the F region, being very sparse retains a good deal of its charge (separate electrons, atoms) because it is hard for the electrons to find an ion to run into in the time allotted by night.

The ionosphere then is an area of active electrical charge that waxes and wanes with exposure to the Sun. Energy can be extracted from this charged region by objects passing through at high speed, such as satellites in orbit. NASA has experimented with conductive tethers trailed from the Shuttle and from satellites, with mixed results. It is a quirky environment, surges and over voltages have damaged equipment, making it too variable to be a practical source of energy at this time. Our ribbon does not pass through the ionosphere at high speed; it is more like a part of the local neighborhood.

As to why people would be worried about the ionosphere being discharged is a mystery, probably best answered by analysts on an individual basis.

The charge production rate in the ionosphere ranges between 2000 and 6000 q/cm^3/s (q = an electron charge). For an area around the ribbon of 1km x 1km and 500km in vertical extent this relates to 1×10^{25} q/s or 625,000 C/s. With a resistivity of $10^{-4}\Omega$m for carbon nanotubes, a 20-ton capacity ribbon (3 mm^2 cross section) would have a minimum resistance of roughly 5MΩ. For the ribbon to discharge the ionosphere at the same rate as charge is being produced would require a current of 625,000 Amps to flow through the ribbon. To produce this current a voltage difference of ~3 x 10^{11} Volts would be required between Earth and the ionosphere. The measured electric field under thunderclouds just before a lightning strike is 10 – 20 kV/m. If we extend this electric field up to the ionosphere (which does not occur but should be a worst case) we find the static voltage potential would be less than 2 x 10^9 Volts. At this voltage difference with no redistribution of charge in the ionosphere we could discharge an area 100m around the ribbon. Since we have assumed the most conducting ribbon possible (in reality it would probably be down by orders of magnitude due the epoxy sections) and the highest potential difference conceivable it is more likely that only a small volume of centimeters radius would show any affect from the ribbon's presence.

Even without the math we should be able to appraise the situation on an intuitive level. We do not drain the whole State of Texas of its thunderstorms by putting up one lighting rod. When we erect a TV antenna we don't drain away all the radio waves so that our neighbors don't get any. (Don't laugh, an obnoxious neighbor once blamed his poor reception on just such a claim.)

Severed Ribbons

If a ribbon is severed the lower segment will fall back to Earth while the upper portion floats outward. The worst case would be if the countermass breaks off the far end of the ribbon and the entire 100,000km of ribbon falls back to Earth.

Depending on the location of the break, the epoxy used, the dynamics of the fall, etc. the ribbon will re-enter the Earth's atmosphere at a velocity sufficient to heat the ribbon above several hundred degrees Celsius (figure 10.9.1). If the ribbon is designed properly, the epoxy in the ribbon composite will disintegrate at this temperature. This means the ribbon above a certain point will re-enter Earth's atmosphere in small segments of carbon nanotube / epoxy dust, chances are, no one would notice. About 3000kg of 3 square millimeter cross-section ribbon (20 ton capacity) may fall to Earth intact and east of the anchor. Depending on just where the anchor ship happened to be that day, there would be from 1500 to 3000km of open ocean east to the coast of Ecuador so, except for any fish in the top 1 inch of water right on the equator, again, no one would notice. Also this portion would most likely be slowed to "terminal velocity" like any other small object falling in air. In this case the speed of falling leaves. Keep in mind that the ribbon is very light, wider than a full page of a newspaper and less dense. If it were to hit you on the head you might not notice. Falling bed sheets would be more dangerous. Detailed simulations will be required to determine the possible sizes of segments that will survive and the health risks associated with carbon nanotube and epoxy dust. In terms of the mass of dust and debris that will be deposited, we can compare what will happen to what naturally happens now. Each year 10,000 tons of dust accrete onto Earth from space, the additional 890 tons of the first ribbon will increase that year's infall by 8.9%. Larger, 200 to 1000-ton capacity ribbons would have a mass of 8,900 to 44,500 tons or roughly equivalent to 4 years, for the largest, of normal global dust accretion. Further investigations are required to determine the environmental impact of depositing this much dust along the Earth's equator. But consider for comparison that at any given time a good dust storm would deposit more than the ribbon over a smaller region, and in any given year forest fires would distribute so much soot over the land as to make the ribbons dust statistically unnoticeable.

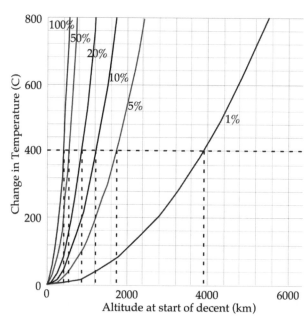

Fig. 10.9.1: Re-entry temperature of the ribbon as a function of starting altitude and fraction of energy deposited in the ribbon. Depending on the fraction of energy that goes into the ribbon (1% - 100% shown in the plot) as opposed to going into the air, ribbon lengths of 400 to 4000 kilometers could re-enter without disintegrating.

In the opposite case where the break is at the bottom of the ribbon, the entire structure would float up away from Earth. The ribbon would remain in orbit with the lower end hovering above the Earth at some low altitude. If a weight of greater than 20,000kg is lowered down on the ribbon and stopped to hang on the lower end of the ribbon the ribbon would float back down to earth and could be reconnected. How this would be done depends on whither the loose end was still in the atmosphere or above it, beyond that we will not speculate on here.

In all cases, there would be no need to run in panic, there will be some amount of time (hours to days) between the initial break and any substantial change in the configuration of the ribbon. Various scenarios can be derived for saving the ribbon or stopping it from re-entering during this delay but these will all greatly depend on many aspects of the design and current state of the ribbon at the time of the break.

In any analysis of the environmental impact the possibility of a falling ribbon and the damage it may cause, the question must be asked, "compared to what?" The alternative, already with us, is the continued use of rockets. During rocket use both pollutants from the burning fuel and from the re-entry of the spent rockets must be considered. For example, each Titan IVB has a dry mass of 65,000kg, much of which ends up re-entering and burning up in Earth's atmosphere. The Titan IVB also burns roughly 500,000kg of propellant. Our proposed 20 ton capacity ribbon has a mass of 890,000kg, one and a half Titans worth. Each Shuttle launch puts 2,000,000kg, of propellant and debris through the atmosphere; so each year the ribbon would save 100 million kg Not being put into the air. A strictly mass comparison is far from the only comparison to make but it gives a rough idea of scales of the environmental impacts we need to consider.

Chapter 11
Budget Estimates

Estimating the cost of building and operating a space elevator has been quite a challenge. At each stage of development, new methods, technologies and materials have had significant effects on the proposed budget, fortunately all in favor of a lower one. Confidence grows that this project can be done for less than the national debt. There are still many unknown areas, but we feel, for at least large segments of the program we can get cost estimates that can guide future decisions. We feel we have good cost estimates for much of the program, the ocean going power and support system being the most realistic and the ribbon production being the largest uncertainty. We have also tried to make conservative cost estimates. The bottom line is that the space elevator could be built at a cost comparable to many U.S. endeavors and a lot less than some that have been proposed.

Launch of the Initial Spacecraft

As we have progressed through this investigation several aspects of our program have changed since the first study. The launch method is one of them. One would think that in such a project as this, producing an unknown object like fabricating the ribbon would be the most costly. And it still may be, but as it turned out, dealing with existing technology, getting the darn thing up into the sky proved to be one of the most troublesome and costly. (What better testimonial that we need a new space launch system?)

As we detailed in chapter 5 there isn't anything large enough to launch the whole assembly to LEO, it has to be sent up in pieces, assembled there and then rocket boosted to GEO where deployment of the ribbon can begin. In our original investigation it looked like we would have to use the shuttle ($500 million each) to get the largest piece, a 22 ton pilot ribbon and its spool, up to LEO, and later another shuttle (another half a billion) with the crew to assemble the pieces. Then there was the cluster of Centaur rockets and their fuel for the second stage trip to GEO. All of this would take another 7 commercial launch vehicles, primarily because the fuel for this second stage was 127 tons! For conventional, off the shelf hardware, the total cost was to be $3.6 billion, and we didn't even get any new technology out of it! Or we could go with the, already on the drawing board, Shuttle-C launch option, 77 ton lift capacity. This would involve an additional $3B - $7B (likely $5B) for development and several more years. But we could save $2B for the five fewer launches and we get a larger first ribbon – a major advantage, later, during the ribbon build up phase. The launch costs on the shuttle-C option will be $4.7B - $8.7B.

What the heck, at those prices let's try the unconventional.

Slow and steady wins the race

Unlike taking off from Earth where speed must be gained as fast as possible, once in space, changing orbits can be done at one's leisure. Theoretically, if not practically, a thrust, however small, if sustained long enough, will get you to your destination. So, we don't have to use the standard liquid fuel engines, we can use the new ion and plasma exhaust engines that are far more efficient. These MPD (magneto-plasma-dynamic) engines use the principle of very high exhaust speed to get more work out of a given mass of fuel. (See chapter 5 for details.) Such engines are not very massive and neither is the fuel, less than 10 tons for transport to GEO as opposed to 127 tons with chemical engines. We gain another substantial savings over what was originally the third stage propulsion system.

In the original plan a third, low thrust, propulsion system was needed for the ribbon deployment phase, where the spacecraft is moved from GEO out to its full extension at 100,000km while the ribbon is being dropped down to earth. That system would have been 2 tons for the engines and 18 tons more fuel. The MPD engines are perfect for this kind of work where on/off and variable thrust operation is needed. The engines are already accounted for (to get to GEO) and the fuel needed for this phase is only two additional tons. This allows for a wonderful bonus, we can afford (the mass) to send up two spools of ribbon and get a pilot ribbon almost twice as large as the original plan. The two ribbons would be deployed together, side by side with only the ends attached at the descending endmass deploy craft. The first climber up the ribbon would then join them, in something like a zipper operation. This saves us tens of millions of dollars and several months time later in the program.

Our revised configuration now consists of two 20 ton ribbons, on two spool/dispensers, 800 kW MPD engines, the spacecraft frame, components and fuel. This all weighs less that 96 tons (half the original plan) and can be carried to LEO on 4 Atlas V Heavy or Delta IV Heavy rockets costing between $200 and $250 million each. Our launch budget is now down to less than $1 billion. Savings, $2.7+ billion.

Impressive savings? Well, we cheated; notice that there is no power source. Chemical rockets are their own power source; you burn the fuel. In electrical engines like the MPD the fuel is only the reaction mass, electrical energy must be applied to throw it overboard at a useful speed.

In this case we left the power source on the ground and we are going to beam it up to the ship via lasers, where it will be converted by photovoltaic panels (solar cells) to electrical power. This saves tremendously on mass (and therefore dollars) that doesn't have to be transported to space and we don't give back these savings by using the fancy lasers since we are building them anyway as the power method to operate the ribbon climbers.

Initial Spacecraft

The initial spacecraft, for the trip from LEO to GEO, would consist of the engines on a mounting platform, fuel tanks and fuel delivery system, control and communication package, attitude control system and its fuel supply, photovoltaic panels and mounting cradle for the payload, counterweight mass and the ribbon deployment components. It will be a relatively simple design basically an open frame, but one of the larger systems ever launched. It would be designed to be partly assembled in orbit – a "plug in" arrangement to minimize hands on labor. The main purpose of the spacecraft is to take its cargo (the ribbon) to GEO orbit and deploy it. There are no complex optics or electronics, but there are mechanical systems. We believe this spacecraft could be built for approximately $300M. See table 11.1 for cost details of the components.

Ribbon Production

This was the most difficult component of the program to get an accurate cost estimate for; a ribbon, such as we are discussing, has never been produced before. Fortunately, the CNT technology is advancing more rapidly than we would have expected. Mitsui now estimates that we will be able to get the basic CNT material at roughly $100/kg. In addition to the two 20 ton ribbons that will be deployed from space, there are 230 additional ribbons which will be added via the climbers. The manufacturing must be a completely automated process from feeding in the raw materials to final testing. The problems are not with the actual manufacturing. The large-scale production of continuous pieces as we are discussing is nothing new to many industries (textile, automotive, fiber optic, electronic, etc.). Rather the major problems are twofold, the production of the nanotube raw material in sufficient quantities and quality, and the actual assembly process; this is more like an oriental rug than a hoist cable. Once these problems are solved, the actual production should go rather well.

Already nanotube production and experimentation is proliferating all over the world. Commercial nanotube products, of various kinds, will be out in the market long before we build our ribbon. Initial discussion with 3M indicate that the cross-ties adhesive material could be developed (and supplied) for as little as $20M from existing materials that are already close to what we need for the ribbon. For quality, and scheduling control we will need a dedicated production facility at a cost between $100M and $150M. All this has allowed us to lower our estimate from the original $5B (the cost of advancing the technology by ourselves) to a more modest $397M.

Climbers

The 230 climbers we intend to send up the ribbon will all be the same basic design but need to deal with the ever-increasing ribbon sizes on increasing climber sizes. The new larger pilot ribbon allows us to start out with larger initial climbers, carrying almost twice as much add-on ribbon as the original plan; the first climber almost 1 ton, the last climber will be 20 ton. Hopefully the climbers can be designed in a modular form so that for some large number of climbers the same components can be used, just in larger quantity. For example, the first 20 climbers may have two motors to pull them up the ribbon. The next 20 may have several structural components and one motor added without substantial modification to the rest of the climber. If designed properly each of the 230 climbers will be a bit larger than the last, without it being a custom production. The climbers are fairly simple systems consisting of photocells for receiving the beamed power, DC electric motors, tread based drive, structures and controls.

The motors will be roughly $50k for each 100kW of drive power at least initially but future production could drop this cost. The photocells, as with our spacecraft, will be $10k/m2. The structure and control will amount to $200k per climber. While not enough units to constitute "mass production" we will get considerable savings from what amounts to long production runs. We will also be building dozens of cargo carrying climbers to operate the ribbon. We now put the first construction climbers at $1.6M each (excluding the ribbon cost), declining with volume to about $0.5M per climber by the time we are well into regular operations with cargo climbers.

New Tracking Facility

We must not allow orbiting objects to impact the space elevator, that is why we have made it mobile so that we can move it out of the way of on-coming objects. That requires that you know the time and trajectory of such objects. Radar systems such as the Haystack Observatory have begun tracking objects down to 1 cm in size. Optical systems, which have some advantages (and disadvantages) are also coming on line. NASA has already requested that the current space surveillance network be upgraded to track debris down to 1 cm with improved accuracy. It has also been suggested the upgrade include moving the facilities to the equator, since everything in orbit must cross the equator, and with less slant range to deal with greater accuracy can be achieved. More details are required, but if the upgrade were to occur, it may satisfy all of the foreseeable tracking needs of the space elevator.

If the upgraded space surveillance system does not come online before a space elevator is undertaken, then a set of tracking facilities (radar similar to Haystack or a phased array or optical as suggested by Ho) would have to be built. A reasonable estimate is that five new facilities would be required along the equator. Five facilities would allow at least three stations to cross reference, increasing accuracy and not place the elevator in jeopardy if one or two tracking facilities were down. Based on estimates for a new radar observatory each facility could cost less than $100M (Berkeley's One Hectare Telescope for example is $25M) including high-speed computing facilities.

Johnson Space Center has also conducted a study on constructing a new set of tracking facilities. The system that was proposed used current technology (X-band phased arrays and dishes) and U.S. locations. The total estimated cost of one billion dollars to build and 100 million dollars per year to operate were given without any breakdown but are roughly close to our estimate above. A fair price would then seem to be in the range of $500M if we were to build these facilities ourselves.

Since these facilities would service all of the worlds space activities for collision warning and avoidance, one would ordinarily think that it should not be incumbent on us to build it. On the other hand, this service is of such vital importance to our operations that we can't afford downtime that might arise from any number of factors, such as; labor or political disputes, strikes, rate hikes or financial disruption, poor management or late information, etc. Therefore it would be better if we built and operated these facilities. This would raise our operating costs but that would be recoverable by selling the services to other agencies. If space activity explodes, as we expect that it will, this service may well pay for itself. If the new facilities are built by other agencies before we get to it, then we may need to enter into some arrangement that would give us the control that we need, one example would be leasing.

In general, leasing is a viable way to stretch the budget of any project and reduce the up front cash flow. There are several other parts of the system that would be amenable to this form of operation and reduce our front year's budget accordingly. As an example the mortgage or lease equivalent cost of a $500M facility would be about $36M a year.

(Operations of facilities will be broken out with the rest of the program below.)

Anchor Platform

If a platform modeled after the *Sea Launch* facility is selected, the total cost of this system would be on the order of $120M. The only major modifications would be installing a precision tension control system for the ribbon attachment and a climber launch housing. This does not include a power beaming station (see below) or unique new facilities that may be desired such as a nearby, floating airstrip.

Power Beaming Facility

For our cost estimates we will use a laser beaming system. The beaming system will consist of the facility infrastructure, high-power lasers and a large deformable mirror as the beaming device. We will want several power beaming facilities located at the anchor and a separate beaming facility on another platform, somewhat removed, to improve the transmission duty cycle. To insure continuous operation the power beaming system will consist of three ocean-going stations for $100M each based on Sea Launch and Megafloat. Each station will have one or two lasers ($100M each) and an associated set of optics ($125M each) with separate power generation system ($10M each) for the lasers. An option for later is to add an additional system set up on land, something more convenient, like the Mojave desert, to pump climbers at altitude. And so for now, we will budget for six systems to start with, several more will be needed later as traffic develops. This will insure continuous operation even if one or two laser beaming systems are down for maintenance.

The Compower power beaming system is well along in its development and has accurate cost estimates for that system. For instance the lasers for the Berkeley 1MW system are $100M each and the deformable mirrors are $125M each including infrastructure. As you see, we have budgeted at these prices but after we place our first order for half a dozen with many more to be added as traffic builds in following years, they should be far more economical. We can also look forward to savings by the combining of two or more laser beams onto one set of mirror optics. Yet to be fully developed, but looking promising, this technique not only saves on the number of high cost mirror sets to buy, but allows flexibility of power output and greater power per beam for larger climbers.

Administration

At this point, we will include operations for the first seven years of the space elevator project. This should cover the contracting period, to develop the components, launch the pilot ribbon and the build up to the full ribbon for the start of commercial operations. After that we will have to make the money to pay salaries. For 10 years, with 200 people at $150k, with overhead, per year we will budget $30M a year. These people will be scattered around the world coordinating activities. This does not include the personnel associated with the actual research, construction, or equipment operators. Those costs are covered under each such activity. These costs do cover tracking operations, administration, satellite integration, business operations, and costs associated with regulatory compliance. We will eventually need offices in each client country or major commercial user, the FedEx of space transportation.

Facilities

Additional facilities that will be essential are living quarters for the crew and visitors at the anchor station. This will be a separate floating platform with ample space for a small airport, houses and professional facilities. A functional platform at roughly $0.3km^2$ would be about $200M based on the Megafloat system. Additional facilities will be required at various other locations and the cost associated with these will be about $20M.

Miscellaneous and Contingency

For some elements of this program it is extremely difficult to get accurate cost estimates, especially for something as unique as the space elevator which really has no prior completed project for comparison. In addition to contingencies for the specific parts with costing difficulties we have implemented a 30% cost contingency to cover items that have been overlooked or may show-up later in the program and to cover growth in costs. Being conservative by nature it is hard to think of over one Billion dollars as petty cash but that is what contingencies are for.

Summary

The summary of our budget estimate is in table 11.1. This study has clarified the cost of many of the space elevator subsystems, happily resulting in substantial savings over what we thought it might cost in our first proposal. (Political costs are not included.)

Table 11.1: Budget Summery		units	$/unit	total
			Millions of $	
Launch Costs	Delta IV Heavy 200	4	200	800
Spacecraft	MPD propulsion system	4	10	40
	Power receiver photo cells	150	0.01	1.5
	Assembly	1	7	7
	Spacecraft structures, R&D	1	15	15
	Ribbon deployment spools	2	25	50
	Endmass deploy craft	1	20	20
	Control hardware	1	50	50
	Motors, misc hardware	1	10	10
	Fuel tank	1	1	1
	All up assembly R&D	1	100	100
	Subtotal			294.5
Ribbon	CNT (tons)	890	0.1	89
	Production facilities	1	100	100
	Composite prod. Facilities	1	20	20
	Ribbon contingency, 90%	.9	209	188
	Subtotal			397
Climbers	Climber, frame & controls	230	0.2	46
	Motor	1150	0.05	57.5
	Solar cells (m2)	22,000	0.01	220
	Construct	230	0.2	46
	Subtotal			369.5
Power beaming	Ocean-going stations	3	100	300
	Lasers, each station	6	100	600
	Optics, each station	6	125	750
	Power system, each station	3	10	30
	Subtotal			1680
Anchor system	Ship	1	100	100
	Ribbon tension sys.	1	20	20
	Subtotal			120
Tracking	Radar and optical systems	1	500	500
Facilities	Floating platform	1	200	200
	Shore offices	1	20	20
Admin	Gen. operations (FTE 10 years)	2000	0.15	300
Contingency, 30%		0.3	4681	1404
	TOTAL			6085

Optional Additions

Several additions can be made to the overall space elevator program that would be of considerable benefit which include:

1. A dedicated ship for transporting personnel, supplies, equipment, climbers, etc. to and from the mainland (figure 11.1). This may be viewed as more of a necessity and therefore an item that should be in the budget but in reality commercial shippers and transport vessels can be contracted for this work as needed.
2. High-tech development, design, construction and repair facilities located on-board the anchor or nearby platform. Even after the first ribbon is finished we will still be in an R&D mode. A facility such as this will be useful in the continuing research to understand the dynamics of the ribbon and climbers and improving their design. Also in unpredicted or unusual circumstances such as a climber that has been damaged in delivery, dealing with a seized climber, inspecting and repairing the ribbon and the ever-present task of solving problems that arise.

Fig. 11.1: Picture of the Sea Launch dedicated ship Sea Launch Commander used for transport of personnel, launcher, and supplies and for controlling the launch operations. A similar vessel would be valuable in the space elevator program for delivery of personnel, equipment and supplies.

3. Future large-scale facilities. As future ribbons grow to transport million kg payloads and people, much larger facilities will be required. What has been discussed here are only facilities for the first

ribbon. Facilities that should be considered in the future include a megafloat-type platform for personnel in-transit, large staging facilities and facilities for supporting a long-term population.

Beyond the first ribbon

After the first ribbon is complete, the second and following ribbons will be considerably easier, faster and cheaper to build per kilogram of capacity. Climbers can be sent up the first ribbon deploying and building the second ribbon as they climb; only now they start out with 13 tons of new ribbon on each run. In this case a ribbon of comparable size to the original can be made in 7 months instead of the 25 months of the first. A large amount of the cost of the first elevator is the difficulty of starting up production processes that have never been done before. With that behind us and with the elimination of non-recurring costs we find the second ribbon can be built for under $2B (table 11.2), and the third and subsequent ribbons will, in turn, cost less than the second.

Table 11.2: Cost of Producing the Second Ribbon

Component	First Ribbon ($M)	Second Ribbon ($M)
Launch cost to GEO	800	0
Spacecraft	295	0
Ribbon production	397	100
Climbers	369	80
Power beaming stations	1680	1400
Anchor station	120	120
Tracking facility	500	0
Facilities	220	30
Per year operations	30	40
Misc. and contingency	1404	100
TOTAL	~6085	~1870

The value of these ribbons are no less however and, if it were so decided, some of these subsequent ribbons could be sold as completed units to private industry or other entities to pay off our construction loans early. Without the overhead costs of debt service our margins would be fabulous and our competitive position assured. One other cost will have to be considered when the time comes, and that is the loss of revenue while the first ribbon is occupied building the second ribbon. Perhaps the traffic can be interleaved to minimize long down times. However it is handled it will be worth it. With two ribbons there will be backup as well as having, from then on, one or more ribbons for traffic and one or more ribbons for building. Self-sustainable and likely self-financed expansion will be under way. When considering larger ribbons the scaling of the climbers and power beaming systems must be taken into account. For a more detailed discussion of the costs and size of future ribbons and the economics of operating systems of space elevators see chapter 13.

Chapter 12
Schedule

Future work

With the completion of initial studies we find no reasons a space elevator can not be built and find many reasons to build it. We have examined all aspects of the space elevator but there is obviously plenty of work yet to be done before a complete functioning space elevator can be built. Everything from spacecraft structures and motors to the weather need to be examined further.

Of these follow-up studies the most critical ones are:

Nanotube production: Can nanotubes be produced with the required properties? How easily can nanotube production be scaled up?

Small-scale ribbon design: What is the best design for the ribbon on small scales (microns to meters)? We are addressing some of these issues in our ongoing work and hope to have short ribbon segments for testing in the near future. Our work will investigate the interfacial adhesion of the carbon nanotubes and matrix material and production of composites with aligned nanotubes.

Ribbon production: What are the difficulties in producing the ribbon and at the rates required?

Full-up power beaming test: How well does the complete system work at the required power levels?

Ribbon splicing on-orbit: What is the best method to attach additional small fibers to the edge of the initial ribbon?

Damage by small, grazing-incidence meteors: High-velocity impacts tests are required to determine the seriousness of grazing incident meteors. All meteor sizes from 10 microns to 5 millimeters and all impact angles should be tested on different ribbon designs. Our current program should address these questions in the coming year. We will be producing short ribbon segments and placing these in high-velocity impact chambers.

Protection from Atomic Oxygen: Are carbon nanotubes susceptible to atomic oxygen erosion? What are the optimal coatings for protection against atomic oxygen and can they be coated onto carbon nanotubes and epoxy? What is the wear rate of this coating under high-speed climber traffic? Like the meteor and ribbon production work listed above our current program should address the atomic oxygen question as well. We plan to use the ribbon segments we will make and test their resiliency to atomic oxygen in a test facility both with and without a metal protection coating.

In addition our current work is studying the long term weather system at the anchor location for both ambient and extremes of conditions that we might expect, the mathematics of the ribbon (static and dynamic), the toxicology of carbon nanotubes, the modeling of the stresses experienced by the ribbon, the interactions of the anchor and power beaming systems, wear and tear tests on the ribbon, designing the splicing system, refining the budget and schedule numbers, etc. In short, we are hitting everything that is a concern or is still lacking in detail.

One thing that would help in the development and understanding of the space elevator is the successful completion of a few feasibility tests. One such test would entail the use of a high altitude balloon along with a prototype climber and power beaming system as seen in figure 12.1. This feasibility test would demonstrate the operation of several of the primary components of the space elevator together. The test would also show, early on, where additional development is required. (In testing hardware when things don't work they obviously need "further development".) However, this test would not address the deployment or much of the anchoring aspects of the design or the difficulties that will be encountered due to the space environment. This system is also very inexpensive compared to the space testing of individual components that will have to come later and where "all up" or combined testing is difficult.

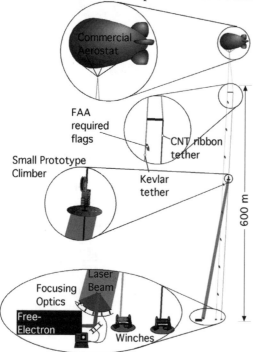

Fig. 12.1: Schematic of the proposed high altitude balloon feasibility test for examining parts of the space elevator system.

This type of test rig is very adaptable and scalable. For instance you can start with any kind of ribbon, steel, Kevlar, whatever, and just test the power beaming and power receiving components. Then you can come back and add the different climber motors, then make a facsimile of our ribbon out of Kevlar or graphite whiskers and test traction rollers, etc. Or you could take things in reverse of the order just mentioned and run power to the unit with plain old wire, if the power beam unit is not ready. The point being, in this way we can piece together a tremendous amount of practical engineering knowledge without being delayed or "held hostage" by the more elusive elements that may take a long development cycle.

Schedule

Based on our proposed design we have estimated a realistic schedule for deployment of a space elevator once the critical technology development efforts are complete. <u>This schedule is based on the technical aspects of the program only</u>. Some of the baseline time constants used to produce this schedule include:

- Time to produce the first 20,000kg capacity ribbon, from the time of launch: 30 months
- Time to double the size of the ribbon:167 days
- Time to upgrade a 20,000kg cap. ribbon to a 200,000kg cap. ribbon:20 months
- Time to upgrade a 20,000kg cap. ribbon to a 10^6kg cap. ribbon:32 months
- Time between climbers: 3 days

This schedule begins with the official start of a space elevator project. That means the money and resource allocations have been approved. Which also means that the preliminary research and development has been successful, showing that a nanotube ribbon can be manufactured and that climbers can climb while splicing ribbon. The time needed to accomplish these preliminary steps is unknown, but we should not be surprised if it takes 5 to 10 years and billions of dollars in primary research. It will depend entirely on whether the development research is a determined, well-funded effort or only advances piecemeal as a sideline to various commercial interests in other nanotube product development. Only then can we put together a viable project.

In addition, we have laid out one possible scenario for future utilization of the space elevator. We have assumed that after deployment of the first several space elevators the system designs, materials, construction techniques, and deployment techniques will improve, dramatically reducing the time and cost of future elevators. We can have confidence in saying this because although the first elevator may be difficult and expensive there is nothing in this technology (after knowing how) that would come even close to the level of sophistication and detail work that some of our current industries take for granted (such as the computer chip industry). Twenty years ago we could not have had one Pentium chip even for $100 billion. It took the first chip and all the others in succession to make chips by the billions, not for a billion. For a more detailed view of the expansion and cost structure of the space elevator system see chapter 13.

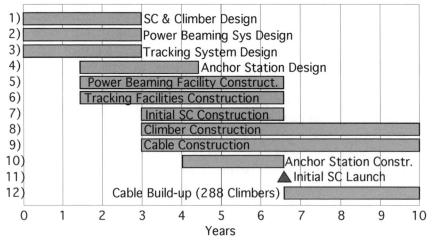

Fig. 12.1: Deployment schedule for a space elevator.

Notes.

1. Spacecraft and Climber Design: The design is always the most critical stage of any space program. The three years that we are allowing here should actually be the final design stage. Again, this is starting after all the technologies have been worked out and we are assuming here that people will do lots of thinking and testing of all the real world, out of the laboratory, aspects of the space elevator before the program begins.

2. Power Beaming System Design: This should be an easy schedule to meet; only needing a final design stage based on the currently available systems. All technology development needs to be completed prior to this stage.

3. Tracking System Design: Since the proposed system is based on current systems this should be a simple schedule target. We are

basically going to add the software oriented on tracking and warning for the ribbon.

4. <u>Anchor Station Design</u>: Based on the *Sea Launch* program, this schedule should provide plenty of time for a proper design to be completed. For the first unit we don't have to design the ship, just the ribbon mounting and climber handling mechanisms.

5. <u>Power Beaming Facility Construction</u>: Based on the commercial Compower system, this should be a very reasonable schedule to build and test the power beaming system. Once again, this is a critical system that will have backups because a power beaming system must operate continuously for the entire time during the climbing stage.

6. <u>Tracking Facilities Construction</u>: Based on current technology and systems, five years for construction should be feasible. The tracking system must be up and have a complete database of objects prior to launch. This again is a critical system that must be operational during the entire life of the ribbon. In fact, since a system adequate to our needs should be up and running, in service to the rest of the space program, we are hoping to remove this item from our budget.

7. <u>Initial Spacecraft Construction</u>: After a detailed design process building the initial spacecraft should be feasible within this time period.

8. <u>Climber Construction</u>: This schedule item extends well beyond the launch of the ribbon because it includes all of the climbers, only the first dozen or so must be ready by launch time. The climber design and facilities to produce them must be such that the entire 230 climbers can be delivered during this seven year period of testing and deployment plus spares for the few malfunctioning units. If the climbers can not be produced during this period, there is a substantial risk of program failure (the ribbon coming down). This point emphasizes the need for a streamlined program and designs that are expandable yet simple.

9. <u>Ribbon Construction</u>: Again, this schedule item extends to the end of ribbon build-up, it includes construction of the initial ribbon and all subsequent ribbons to be taken up the elevator. This item has the same cautionary note as the climber construction above. Each of the 230 ribbons must be produced with almost perfect quality and on schedule, which means keeping many ribbons ahead of the current climber.

10. <u>Anchor Station Construction</u>: The *Sea Launch* platform was constructed in 18 months. Combined with the design stage the time we have allotted will be sufficient to build and test the anchor platform.

11. Initial Spacecraft Launch: This includes launching the various
    spacecraft components and the quick on-orbit assembly. This whole
    process should be designed to take no more than a few months. This
    will require streamlining the launches and possibly utilizing more than
    one type of launch vehicle.

12. Ribbon Build-Up: The first year of this is the most critical. After the
    first year, the ribbon is more stable and less vulnerable to damage by
    meteors.

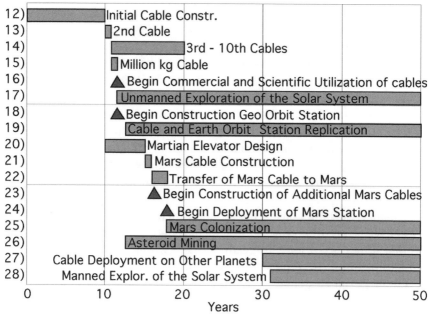

*Fig. 12.2: Deployment schedule and possible long-term utilization
scenario*

Long-term (see figure 12.2)

13. Second Ribbon: Once the first ribbon is in place, building additional
    ribbons becomes much simpler, less risky and less expensive (see
    chapter 13). It would take roughly 210 days to put the second ribbon
    in place based on our analysis. A second ribbon would need its own
    anchor station and power beaming facilities that could be constructed
    during the first climbing stage (#6). The importance of building a
    second ribbon immediately is several fold. First it gives a redundant
    system in case a problem occurs with the original ribbon. Second, the
    two ribbons are completely independent and can address different
    program directions. One ribbon can be producing additional ribbons

while the other is lifting commercial satellites or building up to a much larger size to address manned occupation of space.

14. Third through Tenth Ribbons: A good strategy would be to use one of the first two ribbons to start producing additional ribbons. In 210 days one of the two original ribbons could produce a second identical ribbon. The process can be repeated with both ribbons as often as is needed. For example, utilizing one of the initial two ribbons to produce additional ribbons we could have a total of eight ribbons in 500 days (much less than the nine years we have allotted in the schedule). A point to remember is each of these would require their own anchor and power beaming systems. The advantage of producing multiple ribbons is they can be dedicated or sold to specific users (military, commercial, foreign, and private) to bring in revenue and cover the initial development costs or utilized for different specific tasks.

15. Million Kilogram Ribbon: The larger ribbons will be essential to any manned activities. The larger ribbons will also further reduce the difficulty in producing additional ribbons (a 20,000kg capacity ribbon could be produced with two climbers on the larger ribbon). This schedule estimate of 32 months is based on continuously running climbers up the initial ribbon.

16. Begin Commercial and Scientific Utilization of Ribbons: At this point, the ribbons are stable and redundant ribbons are in place. What this means is that the ribbons will remain in orbit for many decades with minimal maintenance (anchor maneuvers). This is the beginning of the real use of the space elevator. Commercial satellites can be taken to orbit every few days and much larger unmanned scientific missions can be deployed in Earth orbit and to the other planets.

17. Unmanned Exploration of the Solar System with Large Spacecraft: In the current situation, all planetary missions are mass constrained, ten's of millions are spent on making them small enough, in addition to the cost of performing their tasks. The space elevator would allow for very large probes to be sent to all parts of the solar system for detailed and long-term studies.

18. Begin Construction of Geosynchronous Orbit Station: With a large ribbon in place (two is preferable) construction of a human station at geosynchronous orbit can begin. Each climber will be roughly the size of the shuttle orbiter so a station can be built quickly and easily. Depending on the size, complexity and preplanning for the station, the time to habitability could be as little as a month. After construction on Earth a large station (70 times the mass of the shuttle orbiter with 20 million kgs of consumables) with hundreds of permanent

occupants could be placed in orbit in a year (90 climbers). A second large ribbon is preferred for downward transport.

19. Ribbon and Earth Orbit Station Replication: Once the space elevator system is established many ribbons and geosynchronous stations will be constructed. This will be a permanent activity.

20. Martian Elevator Design: A prime example of the utility of the elevator is in manned exploration and colonization of Mars. The design of the Martian elevator is very similar to an Earth one and should be straight forward at this point in the program. The schedule time we have allotted should be more than sufficient. We should also point out here we have left a quiet period of several years during which the Earth ribbon will be utilized for construction of stations, lifting satellites and basically getting accustomed to using the elevator.

21. Mars Elevator Construction: Almost all of the construction time is on the ground with construction of the ribbon, spacecraft and ground module. Assembling the Martian elevator utilizing the Earth elevator and lifting it to Mars will be fairly quick.

22. Transfer of Mars Ribbon to Mars: Travel, orbit insertion and deployment time.

23. Begin Construction and Deployment of Additional Mars Ribbons: By the same arguments as above it is best to have redundant ribbons. Each of these will be exactly identical to the first.

24. Begin Deployment of Mars Station: After the initial or first two ribbons have successfully deployed on Mars a manned module can be sent to Mars. Part of the module will remain in Mars synchronous orbit attached to the ribbon and part will drop down the ribbon to the Martian surface. A return module will be used to complete the journey.

25. Mars Colonization: We have this beginning with the first Mars station deployment but in reality it is easily seen how the system we have described can be replicated many times to basically allow for large scale colonization of Mars.

26. Asteroid Mining: Resource utilization including mining asteroids, tapping power in space and sending it down to Earth, microgravity industries, etc. will become permanent activities.

27. Ribbon Deployment on Other Planets: Earth and Mars are the most ideal planets for establishing ribbons from our perspective. Once the technology is established, the outer moons, Callisto and Ganymede are the next most likely candidates as well as a free spinning colony.

28. Manned Exploration of the Solar System: Once a ribbon is established at a location such as Ganymede, it is a matter of determination to establish a human presence. The timeline on this is obviously a very

rough guess but depending on the human determination, it could happen on the time scale we present here.

For a detailed analysis and exploration of these topics, see the chapters that follow.

Looking at this schedule there is an important fact to realize. The slow part of the schedule is at the beginning if we assume adequate funding is available. Let's consider two roughly equal entities (governments, private enterprise etc.). At year zero, entity one begins building a space elevator behind closed doors. The second is looking at building a space elevator and thinks it is important but has not begun building it yet. At year five the news gets out that the first entity is building the space elevator. The second now jumps into its program and starts building. At year ten the first entity has its first elevator operating and the second entity is 18 months from launch of its initial spacecraft. At year fifteen the first entity has six ribbons up including two 10^6 kg ribbons, has a manned station at geosynchronous, has recouped much of the construction cost through selling two ribbons and through hundreds of lifts on its eight ribbons, and is beginning construction of a Mars ribbon. The second entity has put up its first ribbon. Note that two additional entities also have ribbons now because of entity one's sales. At year twenty, entity one is making billions from the tens of ribbons it has produced, has a manned station on Mars, has a hotel at Geo station which now has a permanent population of over one hundred. Entities two, three, four, five,... each own a handful of ribbons and are trying to compete with entity one. This may sound a bit fanciful but if the estimate of a possible schedule (figure 12.2) is correct then this particular scenario could occur. The scenario of events comes directly from figure 12.2. The point of this discussion is to illustrate the unique advantage of even a single ribbon. With the capabilities of the space elevator, development of space could occur surprisingly fast, with the lead being held by the entity that puts up the first ribbon, all else being roughly equal.

Chapter 13:
Economics of the Space Elevator

Assuming that the research gets done and the first ribbon gets built, is it worth $6 billion, will it pay for itself? And how do you measure worth anyway? Will it make a return on the investment? There are several parts to that. Will the income cover 1) the cost of operations; 2) the initial cost of construction; and 3) future costs of expansion?

It used to be that we didn't have to answer this question, the intuitive answer was that the advancement of science was its own answer. The Apollo program cost $125 billion in today's dollars, over 11 years, and we got back 383kg of rocks. That is $326,000 per gram ($65,200 per carat for the romantic types), pretty expensive rock. Was it worth it? In the political context of the times, sure it was. But from the engineering point of view, of the long term development of space, it wasn't the right way to go about it and it led to a dead spot in our space program. You will notice that we haven't returned to the Moon since. [The right way, in the opinion of many scientists and engineers, would have been to build a manned space station first. From there, you can then send any number of outbound missions, to anywhere, much more economically. For the benefit of those of you born since Apollo, the reason we didn't do it is that we couldn't afford the delay, we had to beat the USSR to dominate space.] None the less, that space program did lead to the development of what is now a continuous commercial use of space, while the non-profit research of space still limps along.

We have come to accept that space is expensive, and despite 40 years of serious effort, counting from Kennedy's speech, it doesn't show much sign of getting appreciably cheaper. This is frustrating because it flies in the face of our common experience of technology during the last 100 years, steadily making things cheaper, better or both. Inflation masks a good deal of this effect. But if you think in terms of what it cost, in time and money, to get oneself from coast to coast 100 years ago compared to now, and then you make the comparison in hours worked to afford the trip you will see the point.

All other industries that one can think of, (trains, airplanes, electronics, chemistry etc.) were able to be developed in a piecemeal fashion. Small beginnings, many by individual inventors (Bell, Edison), new physics, theory, and math by individual geniuses (Maxwell, Einstein, Plank) could be, and often were, done in the kitchen, den or garage on miniscule personal budgets. From these beginnings industry could do useful things on small scales that sustained them long enough to do bigger and better things on large scales. The twentieth century paid for itself from a shoestring to over one trillion dollars a year in just computer technology alone, by piecemeal development. The one notable exception to this was atomic energy. It did need a big budget kick-start and it got that in the atom bomb project of WWII. Two billion dollars later, (22 billion in today's dollars) plus the first couple of power plants and we had ourselves an atomic energy industry. After that initial high-cost push over the early obstacles even this technology became inexpensive, just like our experiences with all other technologies. (Atomic energy is cheap; it is only the paperwork that makes it seem expensive.)

Space, once you get past Goddard, can't be done piecemeal. No Estes rocket is ever going to grow up to deliver a useful payload to LEO. And space doesn't seem to be responding to the big budget kick-start either. After more than half a trillion dollars spent on development, the basic costs of transportation are about the same. Despite this, we do have a space industry because space is so valuable that it is able to sustain a growing use from just the most productive functions.

There are two basic reasons for this cost obstinacy, the large amounts of energy needed to get to any orbit and the lack of mass production of equipment; virtually everything is hand built. (Europeans still build some of their exotic cars partly by hand, and they run from $250,000 to $650,000 each.) The answer then is obvious. We have to use a lot less energy and do it more often. In order for space travel to be truly successful, that is, an everyday resource, as the wheat fields of Kansas are to the bread on your table, the cost of space travel will have to become: an incidental part of the overall cost of what we are trying to get done.

We have, in fact, made progress in using energy more efficiently, the cost per kilogram to LEO has decreased from about $25,000 per kg in the days of Apollo to $4,500 for some commercial launchers. This is not yet an order of magnitude difference, when what we need is more like two to three orders of magnitude improvement to really be able to declare victory. It does not look like there is any way to improve the efficiency of chemical fuels technology enough to give us that level of savings.

The Space Elevator (SE) solves both of these fundamental problems. It uses far less energy to start with and if developed into a multi-unit system could increase the frequency of use by three or more orders of magnitude, thus bringing mass production of hardware into play.

If the first SE costs $6 billion, then:

Without getting fancy, and including the cost of money factor, we can take as the simplest approximation of construction overhead cost, $6 billion divided by 20 years depreciation (it will last longer, this is just for finance purposes), to get a $300 million a year construction overhead cost. Per day charges would be $822,000.

The first years operating costs are included in the starting budget but when considering other years we should use here as an ongoing operating cost of $168,000 per day, bring the daily charge to fee base to $1 million. This is the minimum that we have to recover from operations (the fee base) to make the ribbon "worth it" from a financial point of view.

Operations

The first ribbon, having a capacity of 20 tons, has climbers that weigh 7 tons empty and a payload of 13 tons. At this level of ribbon development, we are using traction type climbers; they have a friction contact. This limits the speed of any vehicle on the ribbon. The upper limit for any traction vehicle, in repeatable operation, that is, not trying to set a record, is about 400km/hr, and the wear factor is high. Therefore, until we find some as yet unknown advantage, we set the practical limit for this level of development at 200km/hr. This makes the standard trip to GEO, 36,000km divided by 200, 180 hours or 7.5 days.

A fully loaded climber uses 2.4 MW (megawatts) of beamed power to run up the ribbon, but with 30% efficiency, the total ("wall plug") power that we have to pay for is closer to 8MW. At 200km/hr, a trip to GEO takes about 180 hours, that comes to 1,440,000 kWh. But, since gravity diminishes with altitude, we do less work per hour as we ascend, so we don't need full power for the whole trip. We still have the same mass (inertia) to push around so as a rough rule of thumb let's say we use half of the kilowatts, 720,000 kWh. At 5 cents a kWh (no telling what it will be by the time we do it) that is $36,000 or $2.77 per kg for the 13,000kg payload, all the way to GEO. So far so good, that compares very favorably with the $50 per payload kg for space shuttle fuel, just to LEO, and over $210 per kg to GEO (that is just comparing "fuel" costs, or energy used, don't confuse with total costs). That's how we satisfy the first requirement of using less energy. A little increase in the efficiency of the system over the years and we will have a full two orders of magnitude reduction in energy usage.

Here is the operating profile just described for the first SE:

Lift capacity total / payload, 20 tons / 13 tons (metric)
Trip to GEO   180 hr. (7.5 days)
Round trip time   360 hr. (15 days)
Climbers, cost $1 million @
Initial cost $6 Billion
Construction OverHead $300 Million/per yr, $822,000/per day
Current Op's cost $65 million/per year, $168,000/per day
The daily charge to fee base is $1 million.

The conventional way to run a hoist (and the only way to launch a shuttle) is to send it up, unload, and then bring it back down. If it takes 7.5 days to get to GEO then it is going to take 7.5 days to come back down, 15 days before we send up another load. That's 24 trips per year, or a 15 day "duty cycle", as they would call it in industry. Each trip would then cost $15 million. Now this is notably better than the Shuttle at 6 to 8 trips per year and up to $500 million per trip. The per kilogram cost is also better, at **$1154**/kg to GEO compared to about $60,000/kg for the shuttle or about $15,000/kg by commercial launch. You will notice that this is about the same as the hoped for costs of the SSTO that *might* be here in 10 years. But still, that's no way to run a railroad when trying to pay off a big loan. We should be able to improve on this schedule by applying a little modern "packing theory".

You don't have to wait for the first ship (climber) to get all the way up the ribbon and back down. You can have more than one climber on the ribbon at one time without exceeding its load limit. The force of gravity decreases quite rapidly with altitude, and therefore so does the "weight" of the climber on the ribbon. At about 14,000km the gravity gradient is down to 10% of Earth surface gravity, or 0.1g, so that climber is now adding less than 2 tons stress on the ribbon. [For you number crunchers, the stand alone (without a ribbon to hang on to) 10% point is 13,800km and the actual altitude on the ribbon is 12,821km if you include the centripetal force effect of the ribbons rotational velocity. But, we will give it a little margin and as you will see it is easier to work with round numbers.]

After 70 hours the first ribbon-ship will be beyond the 14,000km, 0.1g (gravity) level. There now being just a small load on the ribbon you can start up another climber. When the second climber reaches the 0.1g point the first climber will be at 28,000km, or 0.3 tons "weight", the second at 2 tons, so, you can start a third climber. (And when they get near to GEO they will have no "weight" load on the ribbon.) Now you have three climbers on the ribbon. A brief stop to check with our accountant in his office finds him happy as a clam at high tide. As he looks out his corner office window he sees the fourth climber just departing, he now has four climbers over which he can spread the daily $1 million charges. The interval between climber departures has been 70 hours. Each climber only needs to be charged for this interval of time not the whole length of the trip. 70 hours is 2.917 days so the charge is $2.917 million per climber. This lowers the cost per kilogram from $1154 to $224/kg.

Sad it is that we have to tell the dear man that he can't count his kilograms just yet, there is another factor before he can assign charges to the climbers. It is obvious that if we keep this up we are going to have a traffic jam at GEO. We have to either stop sending up ships so that we can bring them all back down or we have to dispose of them somehow. Disposing of climbers certainly grates against the nerves of frugality, to say nothing of the sense of loss for anyone who appreciates beautiful machinery, but let's at least look at it so we know what choices we face.

The "Bic" disposable climbers.

Empty climbers mass 7 tons each, so with a little separation between them we could bring down three climbers as a cluster, cost, no paying travel on the ribbon for 7.5 days. So, if the cost of a climber is less than the product of, $1 million dollars, times the average days lost to bring them down, it would be cheaper to kiss them a fond goodbye. Ironically, the more we throw away the more we manufacture and the cheaper they get per climber, making it easier to throw them away! In this example we can bring down three close together, but 7.5 days of dead time, times $1 million a day is $7.5 million. So, at this stage, if climbers are below $2.5 million each, (which they should be) it would be cheaper to throw them away, and keep on sending up climbers every 70 hours. By the time we are into steady commercial operations the cost of climbers should be well below the $2.5 million throw, no throw, point; more likely down to the $1 million we used in the above profile. The duty cycle is now *really* 70 hours, and the cost per trip is $3.92 million, including the cost of the climber or **$301/kg**. But we can do better.

Traffic management

First we have to know that the 7.5 day trip to GEO is only one of many destinations (or altitudes to go to) on the ribbon; LEO's (various), GEO and several places beyond GEO for sling shot passage to the planets. To send satellites to LEO's of various altitudes, we can drop off the load on the way up, well before GEO. We then have an empty climber (7 tons) that will allow it to have other climbers near it without violating our weight limits. The 0.1g limit, we just talked about happens just before 14,000km. With these two items in mind, we can now try out various schedules.

We start off climbers, c1 through c4 as before, 70 hours, and 14,000km apart. Technically speaking, we are slightly over our weight limit with this method, but in less than 2 hours (at 346km, the "weight" of a climber is down by 10%) we won't be, so we will do it anyway. [If c1 drops its load at the 0.1g point it would help but it doesn't have to, to make this work. In real life this would not be a problem. The best combinations of payload mass (they don't all have to be 13 tons), and climber mass (they don't all have to all be 7 tons empty), would be chosen to allow multiple climbers spaced out at various altitudes along the ribbon.] To continue, in 280 hours we have c_4 at 14,000km, c_3 at 28,000km, with c_2 and c_1 finished, at GEO. At this point we do NOT start c_5. C4 now drops its load (to enter LEO), all other loads have been delivered, somewhere, and all 4 of them start back down. Being empty, even 4 or 5 can travel together except near the bottom where groups of 3 would be the limit. It takes 180 hours for the last one, c_1 to clear the ribbon, and we can start the schedule all over again.

That 180 lost time is 7.5 days to retrieve 4 climbers or 1.875 days of overhead charges to each climber, times $1 million per day equals $1.875 million extra return charge per climber trip. If climbers are only costing us $1 million each it is still cheaper to sling-shot them into space than stop operations and bring them down. There is one other arrangement we can try. We can schedule our clients in a beneficial way. If we make the first trip, c_1, a GEO customer and the next 4 or 5 trips LEO drops at 24,000 or below, we can start c_1 and c_2 down while c_4 is on the way up. Assuming that c_3 was a LEO drop from 24,000km and c_4 a LEO drop from 14,000km, c_1 and c_2 could start back down during part of c_3's trip and all of c_4's trip. This would put all of them back down at 14,000km when c_4 gets there and makes its drop. Spacing them out a little (1200km) it only takes 76 hours to clear down the ribbon. Now the accounting is 76 extra hours for four climbers, equals 19 hours of extra retrieval charge or $792,000. That is less than the $1 million the climbers cost, so we can keep the climbers!

Celebrate with a soft drink, this is only a partial reprieve. It is easy to see demand building up to several hundred trips a year and if we built that many climbers a year their cost could drop well below $1 million, making it very cost effective to make them disposable once again. This is a shame because what if you really liked a particular climber? Our current example is now at 70 hours plus 19 for retrieval for 89 hours charged per trip, equals $3.71 million or **$285/** kg. Even if the climbers cost nothing the cost per kilogram would still be $224 so the best we can hope for with this scenario of cheap disposable climbers is about **$250**/kg.

[I have skipped over the fact that some of the down trips could have paid cargo which could pay for the otherwise non-revenue hours of retrieval and make it worth saving all the climbers on the ribbon at the time. This is a complication we will deal with shortly.]

The only other way to lower costs is to spend more money. Where have we heard that one before? But it's true, our sticking point in the above examples is the $1 million daily overhead with so few trips to charge it to. We have to find a way to increase the number of trips faster than we increase the overhead. (This is, of course, a basic principle of wealth creation, the only legitimate way, small businesses grow to be large businesses.) The way to do that is with more and cheaper ribbons, and now that the first ribbon is built we can do that. A better way to look at it is as a facility of several ribbons, that together provide a service. As we said, we need several ribbons anyway; we need more ribbons for redundancy and safety and to build bigger ribbons. Now let's see how that will pay.

Cheaper by the dozen

Here is how the economies of scale should play out.

Table 13.1: Cost of Producing additional Ribbons

Component	First Ribbon	Second Ribbon	Third Ribbon	Fourth Ribbon
Launch cost to GEO	$1,000M	0	0	0
Spacecraft	$587M	0	0	0
Ribbon production	$390M	$150M	$75M	$30M
Climbers	$161M	$80M	$40M	$30M
Power beaming station	$2,100M	$1,600M	$1,000M	$600M
Power gen. station	$40M	$40M	$40M	$30M
Anchor station	$120M	$120M	$100M	$100M
Tracking facility	$36M	0	0	0
Admin Facilities	$202M	0	0	0
Operation	$210M	$30M	$20M	$10M
Misc. & contingency	$1,154M	$280M	$125M	0
TOTAL	~$6B	~$2.3B	~$1.4B	~$0.8B

Notice how helpful it is that the first time costs, like launching to orbit and miscellaneous uncertain costs drop off. The next big savings is in the three new technology items, ribbon, climbers and power beaming that should get more economical, quickly, with volume. The support systems, power generation and anchor stations, will stay about the same, they are old technology we just order off the shelf. They will get a little cheaper with repeat orders but that might be eaten up in ordering larger or more elaborate versions. By the third ribbon we are cranking out climbers and ribbon material like model T fords, there has to be an Economy of Mass Production that kicks in here sometime; or at least an Economy of Continuous Production. (Many items today are not, in the strict sense, mass-produced. Items like large trucks, heavy equipment and all the commercial versions of kitchen equipment, are still relatively inexpensive for their use because of Continuous Production.) The 4th and subsequent basic ribbons of this size should be less than $1 billion each.

Working Unit	Hoped For Cost	Pay Load	Trips per Year retrieve	Trips per Year disposal
Ribbon 1	$6B	13T	98	125
Ribbon 2	$2.3B	13T	98	125
Total	$8.3B		196	250

If we build the second ribbon we will have two independent ribbons providing twice as much service as before but with combined overhead and service costs. The two ribbons now cost $8.3 billion for both and this cuts our overhead for each ribbon to $0.66 million per day, two thirds of our one ribbon operation. The costs per climber trip are similarly reduced by one third, so that we now stand at a little over **$190/kg** with retrieval and a little under $168/kg with disposal, depending on the cost of climbers. We have cut it down another notch, let's keep going.

One way traffic

Our project is as yet young. If we had a manned GEO space station on the ribbon and mature space commerce we could run one ribbon all up and the other all down. At the station the cars would be transferred over to the down ribbon. (One way to get mature space commerce is the bootstrap method, build the space station at GEO and thus create the traffic.) The duty cycle would then be as low as it can go with the weight limits, 70 hours, and the overhead charge is now **$150/kg**, per ribbon. That is 125 paid trips per year, per ribbon, the same as the "disposal" rate in table above, but in this case we get to keep the climbers. This is also for full-length trips to GEO, without figuring any fancy discounts for short drop-off points; now it doesn't matter, the climber goes all the way. But we don't get that rate unless climbers are also earning money going down.

Two way paid traffic gets maximum use out of the equipment, but we won't have fully paid down traffic until there is as much stuff going down as up, meaning, until we are producing or importing things from space, after a considerable period of economic development of space. Some paid down traffic will start as soon as we have man in space, such as the building of the GEO station. From then on a certain percentage of the down traffic can be revenue generating, mostly returning people.

Now we run into the problem that the 20 ton ribbon with its 13 ton payload is not a good size for serving people traffic. Needing a pressure hull and supplies a 13 ton capsule would only accommodate about 4 people. Already we need a bigger ribbon! The second ribbon now has another important purpose, to build more ribbons.

Building larger ribbons

First we should drop back a bit in our story and mention how the second ribbon was built. The climbers on a "20 ton" capacity ribbon, have a payload of 13 tons. Each trip, hanging on the first ribbon, but building the new ribbon right along side of it, can now dispense 13 tons of new ribbon. We don't have to have varying size climbers that get steadily larger; we can work with just one type of climber carrying the same amount of ribbon. This takes only 6 months to make each, new, 20 ton ribbon. The second ribbon now builds the third ribbon in the same way, in 13 ton additions, only now we are going to keep right on building past the 20 ton capacity size.

As soon as we have the third ribbon (working from the second) built up to a little past the 20 ton capacity size we can stop using the second ribbon. Now we can start running the splicing climbers on the third ribbon itself because we can send up larger climbers with more ribbon each trip in the same way we built the first ribbon. Now things start going very quickly.

How large should we go? Well, we must admit to having had our heart set on a 1000 ton capacity ribbon, it would be quite grand, we could call it something like the "million kilogram" ribbon or the "billiogram" ribbon - such a nice ring to it. It would mass 44,535 tons and have climbers with payload capacities in the 700 ton range, larger than 747's. But, after all the discussion that has been stirred up by the first space elevator report some colleagues have pointed out many good reasons why a 1000 ton capacity ribbon is a bit too big, at least for the early stages of elevator development. So, we will have to settle for something more practical for the third ribbon, a 200 ton capacity ribbon ought to do nicely and at least it is a full order of magnitude improvement.

A 200 ton capacity ribbon would mass about 8,907 tons. (All the discussion of larger ribbons assumes that they will be built to the same lengths and limits. But, as we build larger ribbons we most likely will take the opportunity to make longer ribbons thus increasing the free energy throw range to facilitate access to the outer planets.) As mentioned above, once the third ribbon gets past the 20 ton capacity size the splicers start climbing on the third ribbon itself and increase the load with each climber. The last splicer would be just short of the ribbons full capacity of 200 tons and a 140 ton payload of ribbon. It would only take about 20 months to build a 20 ton ribbon up to a 200 ton ribbon. By the time we get to this stage, ordering ribbon by the thousands of tons, climbers and lasers by the gross, we should be well into the savings of mass production, therefore we estimate that the first 200 ton ribbon could cost as little as $5 billion, with accessories.

As we have seen, we need two ribbons of the same size to really get the economics working and we will be adding additional strong arguments for this policy in a moment. So the first thing we do with our 200 ton capacity ribbon is build another one off of it, like we did with the second ribbon off of the first. Economics of scale should still be working and our estimate for this ribbon is down to $4 billion. As can be seen in table 13.1 the ribbon itself, being most susceptible to the application of mass production, is not the leading cost item – it's the power beaming lasers and lenses. They make up over $2/3^{rd}$ of the cost of later ribbons, so as their cost drops with volume so will future ribbons. Will it never end? No, but we can stop now, for a while, at four ribbons and examine our situation. We now have a "facility" of ribbons with very powerful economic advantages. Let's look at the situation in more detail.

Working Unit	Estimated Cost	Payload	Trips per Year
Ribbon 1	$6B	13T	125
Ribbon 2	$2.3B	13T	125
Ribbon 3	$5B	140T	125
Ribbon 4	$4B	140T	125
Total	$17.3B		500

We now have a system of four ribbons.
Initial cost $17.3 Billion
Construction OverHead $865 Million/per yr, $2.37 million/per day
Current Op's OH cost $110 million/per yr, $0.3 million/per day
Direct Op's Cost(Power) $2.77/per kg/climber

Our daily charge to fee base is now $2.67+ million.

If we use the same traffic plan that we last used with the pair of 20 ton ribbons, all up one and down the other, we would have a 70 hour duty cycle for each pair. We have to do an extra step here because we have two different payloads, one 13 tons and the other 140 tons, but combined it works out to **$51/kg**. Now we are getting someplace! This is a serious saving over the 20 ton ribbon operation and well over a second order of magnitude improvement over the shuttle. But, we can do even better!

Fractional load packing (FLP)

A 200 ton climber with a payload of 140 tons is a large object, almost twice as large as the shuttle orbiter. It is enough to lift 80 to120 people at a time or the rare single item that might weigh that much. But now that we have this generous sized accommodation, we find that it works better, most of the time, if we don't use full sized climbers.

A big advantage of the larger ribbons is that the climbers don't have to all be full size to have a respectable carrying capacity. If we use smaller climbers we can run them closer together on the ribbon, less time between starts. It also gives us the advantage of being able to use a variety of climber sizes as traffic demands. All the original 20 ton climbers of our first ribbon can run on the 200 ton, serving all of, what now turns out to be, the small load traffic.

Climbers in the 50 to 66 ton size range are particularly interesting. The 50 ton would have a payload of 35 tons, 15 tons empty. This is a payload about 40% better than the Space Shuttle and once empty they are light enough to be brought down on the small ribbons 6 at a time, (spaced out to keep the "weight" from exceeding 21 tons). Duty cycle for retrieval, 40 hours. This means that we can send a sequence of 12 climbers up the large ribbon and bring them down the 2 small ribbons without interrupting the large ribbon, duty cycle 15 hours. We could keep a continuous string of 24, 50 ton climbers moving on the large ribbon without exceeding its limits (duty cycle 7.5 hours); assuming that they would be coming down on one of the other ribbons.

The 66 ton climber is particularly interesting for a number of reasons. Its empty weight is 20 tons, the limit of what can be brought down on the small ribbons (one at a time, 70 hours apart) keeping more space available for paid traffic on the large ribbons. Its payload is 46 tons, just the range needed for several types of manned missions (Mars, for instance) and heavy lift projects for which, otherwise, the Shuttle C would have to be developed. (So right there, we should be able to take a $5 billion credit against our overhead charges, for not having to pay for the shuttle C.) And 66 is one third of 200, a key ratio which will become important in a moment.

One of the intriguing things about operating in a diminishing gravity-well is that the smaller the load relative to the size of the ribbon the higher the cost efficiency. Or to put it another way, the closer the loads can run on the ribbon the lower the cost per load. One of us (Westling) discovered that there is even a mathematical relationship that describes this. He calls it the rule of Fractional Load Packing (FLP), expressed in number of climbers running a ribbon at one time. "Within an upper limit of 1/3 of the ribbons capacity, the maximum number of climbers is 6C/M, where C is the ribbon capacity and M is the mass of the proposed climber." E.g. on a 200 ton capacity ribbon, 6*200/50 = 24, that is, 24, 50 ton climbers can operate in spaced sequence, duty cycle 7.5 hours; or 18, 66 ton climbers, duty cycle 10 hours. The FLP relationship is also telling us that, in a diminishing gravity well a static line can carry 6 times its rated capacity if the load is distributed properly. Given these relationships, we can create tables of traffic density, capacity, and costs for any size pair of ribbons.

Traffic density & costs:

For a system of four ribbons, two, 20 ton; two, 200 ton:
Overhead: $2.67 million/day, $111,250 /hour

For each, 200 ton capacity ribbon:
Table 13.2

Climbers		Units On Ribbon	Duty Cycle	Overhead Share	Cost Per kg GEO	Cargo /Day	Cargo/Day 4 ribbon Sys	Cargo /year, 4 ribbons
Size	PL							
Tons			Hours	$(1,000)	$	Tons		
20	13	60	3	$151.7	$11.7	104	217	79,174
40	28	30	6	$303.4	$10.8	112	233	85,014
50	35	24	7.5	$379.3	$10.8	112	233	85,014
66	46	18	10	$505.7	$10.8	112	233	85,014

This table assumes, ribbons working in pairs, one way traffic, paid loads both ways to GEO.
[To analyze one ribbon, OH is prorated by capacity, e.g. 200/440. OH = $50,568 / hour]

The first two columns are the climbers size and its payload. The third Column is the density, how many climbers can be on the ribbon at one time without exceeding its weight limit, from the FLP formula. (Remember, the "weight" is "disappearing" as they gain altitude and zero at GEO.) The 4th column, duty cycle, is the hours of departure delay between climbers to keep the weight spacing proper. It is the trip length, 180 hours, divided by number of climbers on the ribbon. The 5th column is each climbers share, or contribution to overhead. If it is 3 hours between departures then that climber gets charged 3 hours of OH. From the OH share we compute column 6, the all-important cost per kilogram; OH share divided by the payload.

The last three columns show the cargo density, what the ribbon can transport per day and what the system of four ribbons would then transport per day and the system per year. Call this the "magic of compound climbers". (Note that the 20 ton has a 65% payload, whereas the others have a 70% payload.)

To these cost we only need to add the cost of energy, which if efficiencies remain linear, should be the same $2.77/kg that we computed previously. This gives this family of ribbons a full traffic basic fee rate of **$13.5/kg**. We have achieved more than our three orders of magnitude savings over the shuttle, and more than one order of magnitude improvement over the first ribbon.

To put this in perspective, with two 200 ton capacity ribbons established we will have a total lift capacity of 116.5 tons of up traffic payload a day, and another 116.5 tons down traffic. Of that, the two 20 ton ribbons contribute 4.5 tons each way; these would probably be used for light loads, maintenance, company business or other support activities. This 116 tons capacity would equal 4.7 Shuttle launches per day, which is the maximum they could now do in 6 months. By that measure, we have just increased our yearly lift capacity by 180 times, for a total operating budget (cost) of less than $1 billion a year!

We can still use climbers larger than 66 tons. The full 200 ton, 140 ton payload climber would be great for moving people and large sections of space hardware when absolutely necessary, but they would go at a premium price because they would clog up the traffic flow, no FLP. Also, they would not be manageable from the type of anchor ships we have been installing. They would require handling cranes as large as the ship-building cranes at Newport News, among the worlds largest. This will have to await the coming of really large support facilities, which we will be discussing in the next phase of development.

Other nice places to visit

Before we get too far into building large ribbons we better say a few things about how we are going to justify larger ribbons. Where will the traffic come from?

There are so many developments in space that would increase ribbon traffic that we will have to devote separate chapters to these subjects where they can be presented in greater detail. Here are the types of activities we will cover.

Space Elevator affect on space hardware development

- LEO traffic
- Beyond GEO
- Real-estate
- Mars ribbons
- Tourist industry
- Large space telescopes
- Outer Solar System

- Building GEO stations
- Solar Power Satellites (SPS)
- Beaming power to space from SPS's
- Moon bases and ^{3}He mining
- Asteroid mining: NEA's, The belt
- Industrial satellites & trade
- Uranus – ^{3}He mining

For now we can say with some confidence that the traffic will be there, in fact, it will probably develop faster than we can build ribbon capacity. With that in mind let's see where our ribbon development will take us.

Incidental expenses

While we have our eyes on the sky let's not forget what's going on down below. With several ribbons up and perhaps more to come we should be accumulating a significant navy milling about in our little area of the Pacific. We will need an anchor ship for each ribbon and the larger the ribbon the larger the ship. In addition we are going to need support ships to house personnel, supplies, equipment, climbers, construction and repair facilities and to transport all of this back and forth to the mainland. Then there are the power beaming systems, of which we will be accumulating dozens, along with their power generating plants. Eventually this will have to be nuclear powered – we will be using electrical power like a small city, heck, we are going to be a small city. And then there is the matter of shipping all the satellites, people and whatever else is destined for, or from, space.

It would sure be nice if we had an airport. As mentioned briefly in the budget chapter, the Japanese are working on just such a thing, a floating airport for use off of their crowded coastlines. These large floating platforms are several kilometers in dimension; the MegaFloat system dissipates the swells and maintains a level runway. We can expand on this idea and build two, another one for all the above support facilities, storage, especially of large objects like 200 ton climbers and personnel just mentioned, that don't have to be right on the mobile anchor ships. If they will work in the waters around Japan they ought to work great in the overly dull weather zone we have chosen. We should have the Japanese build them and tow them across the ocean. These waterside accouterments shouldn't cost more than, oh, several billion.

Towards the optimum ribbon

When you count up all the places to go and things to build it is easy to see that we will soon run out of capacity. As we will see in subsequent chapters, just one project, the solar power satellite program could use up all of our 200 ton ribbon capacity in its first year of operation. Keep in mind that 85,000 tons a year is not a lot of traffic for an industrial center here on Earth. (One supermarket can move 4,000 tons of goods a year.) In fact, including Mars and other places to put ribbons, and counting replacements in 50 to 100 years, we will probably be in the ribbon building business forever. The nice part to look forward to is that each addition to the facilities makes the system more economical. It's not easy to foresee what ribbons we will build next. It sure would be grand to build that million kilo ribbon, and we probably will when it becomes routine to handle 1000 ton objects with ground handling equipment. Except for rare situations, this is something we have no call to do now and would have to develop a separate industry for. More likely we will use the FLP rule and build ribbons to a capacity three times the mass that we can get a steady use out of, so that the ribbon can be fully used by FLP rules.

By this reasoning we might likely pick as our next addition a 166 ton climber able to carry 70 to 100 people or 116 tons of cargo on a pair of 500 ton capacity ribbons at the FLP rate of 18 per ribbon, 10 hour departures. By this time, (we would have already built over 19,600 tons of ribbon and thousands of climbers) prices for nanotube ribbon should be approaching $20 per kg, and for climbers between $25 and $50 thousand per ton, like heavy trucks. This would put 500 ton ribbons at about $6 billion each and the economics of our space elevator shipping system will have improved once again.

Now our daily operating nut is up, from $2.67 to about $5 million, a 84% increase, while our daily capacity is up a lot, from 233 tons to 793 tons, a 240% increase, (a total of 289,000 tons a year). This brings our basic fee down from $13.5/kg to **$9**/kg, including power. This level of capacity is sufficient to support the building of 2 to 3 solar power satellites a year (the world could use 60 to 100 of them) plus the local traffic for LEO and GEO satellites.

Throughout this discussion, in order to keep things simple and focus on the cost relationships we have treated our overhead costs as just an exchange of dollars. This is not like the real world where there would be interest on long term loans, sinking funds for all the extra facilities we are adding on the ground, in space, charges for people traffic consumables, insurance, etc, and a markup for profit. If you prefer you can go back over these per kilogram rates at each stage of expansion and add on a 100% markup and you will have it about covered.* We did not do that with that other large government transportation project, the Interstate Highway system. We sunk the original costs and used the gasoline tax as a, very much trailing, cash flow fund, and look how the traffic grew. In our case we are trying to increase space traffic by two, even three decimal places during the first 20 years of operations, therefore we suggest a pricing system similar to the above. It provides considerable excess over current operating costs, eliminating any need for subsidies in that regard, while still providing the significant savings to users needed to expand usage. There are plenty of profits to be had later in the project simply by shifting pricing policy at a point where retaining additional cost savings as margins no longer has an appreciable effect on traffic volume.

*[Review: on the first ribbon, of the $6 billion, we allotted $300 million a year, over 20 years to construction costs. That would have been $524 million a year if we had to pay interest at 6%. That alone would be a

57% increase. Since all the other ribbon examples were "costed" the same way, by the time you add on insurance, miscellaneous, and profit margin, a 100% fudge factor as suggested seems quite reasonable.]

Once the first two ribbons are paid off and overhead costs are dominated by the more recent, more economical additions to the system, the overall base fee drops quickly to a level that allows room for margin markups and a return of all startup capital. The base fee for just the 500 ton ribbons by themselves is only $2.94/per kg, about the same as the energy cost. At that point of development one would hardly think there would be serious complaints over charges in the $10 to $12/kg range, a margin of 100%. Demand will grow, more ribbons will be built and each addition to the system will lower the overhead factor another notch.

There will be a time, perhaps 20 to 30 years from the start of operations, at which the cost of materials will no longer be exotic. Components such as nanotubes and power lasers will be commodities and the chief costs will be, as it is for all durable goods, in the manufacturing and current operating costs. As a guess, it looks like the costs of construction overhead would asymptotically approach $1.00/kg per trip, over the long run, leaving the largest cost factor the energy supply. Since the direct energy cost with 100% efficiency is only $0.55/kg there is plenty of room from the $2.77/kg we have been using to find ways to deliver power more economically. Onboard power would be ideal, if we can learn to build small inexpensive nuclear reactors we could get about a 100% improvement in efficiency, including the loss from carrying that dead weight around. Then the ultimate price of Space elevator trips, with margins, in current dollars, will approach $3.00/kg. For the distance, lower than any other form of commercial transportation. People will then be able to make round trip space cruises to GEO for about $6,972 not including food and drink service, leisure travel would be well under way.

As we develop commercial operations and the other areas we will be discussing in the following chapters our need for capacity will continue to grow and we will be adding pairs of 500 ton, perhaps even 1000 ton ribbons every 5 to 10 years, indefinitely. We can look forward to space traffic of 10 million tons per year early in the 22nd century.

Synopsis

Here are tables of values for the traffic conditions discussed above.

Table 13.3 Traffic on the 20 ton capacity ribbon(s). (In the order of above discussion.)

Ribbons, Capacity	Fee Base: Cap+OH ($M/day)	$/kg ($)	Duty Cycle (Hours)	Trips /Year	Tons /Year	Disposal? Yes	Cost/ climber ($M)
1, 20T	1	1154	360	24	316		
1, 20T	1	301	70	125	1627	Y	1
1, 20T	1	285	89	98	1280		
1, 20T	1	250	70	125	1627	Y	0.5
2, 20T alone	0.66	190	89	197	1627		
2, 20T+station	0.66	150	70	250	3250		

** Notes: $M = units in Millions dollars. Costs per trip and tons per year are for the payload, not the whole climber. Cost per kilogram for the traffic models with disposal of climbers includes the cost of climber.

This table shows that due to the high initial cost of establishing the first ribbon (column two, capital overhead) it will take good traffic management and even the disposal of climbers to get the costs per trip down. The last rows show the significant effect of adding a second ribbon to cost per trip and to yearly capacity. As with any human endeavor, 'practice makes perfect'. Once we get past the first two small ribbons and start to build larger ribbons costs drop precipitously.

Table 13.4 Future traffic on larger ribbons. (In the order of above discussion.)

Ribbons, Capacity	Fee Base: Cap+OH ($M/day)	$/kg ($)	Duty Cycle (Hours)	Trips /Year	Trips /Day	Tons /Year	
2x20T+2x200T	2.67	51	70	501	59.33	21656	
2+2x200T,FLP	2.67	13.5	10	1949	233	85,000	
2+2+2x500T,FLP	4.9	9	10	3701	793	289,400	
							Retail
1,500T,FLP	0.82	2.94	10	876	280	102,200	$10/kg
Max 500T,FLP	0.31	1.11	10	876	280	102,200	$3/kg

Notes: The first row shows two 20 ton ribbons plus two 200 ton ribbons with the same traffic model as used with the small ribbons. The second row shows the same set with FLP traffic methods. The third row shows two small, two 200 ton and two 500 ton under FLP methods. These three rows show the effect of including the high cost of the first ribbons. The fifth row shows the economy of each added 500 ton ribbon, without the drag of earlier costs. The last line shows the maximum economy, likely, from each additional future ribbon. In the last column we note the probable retail price per kg, per trip, with profit margins, for later ribbons.

The comparison between rows one and two shows the striking difference that FLP management can make in using the same set of ribbons. The last two rows show the dramatic cost leverage available from using larger ribbons in a mature industry. The large ribbons add rapidly to the yearly tonnage.

[The FLP relationship has other quick ratios that can be used in contemplating any future large ribbons. For instance, the rated capacity of the ribbon (i.e. 500 tons) times the constant 0.56 gives the daily traffic capacity of the ribbon in tons, (500 x <u>0.56</u> = 280 tons/day.]

Chapter 14:
Other Nice Places To Visit

What would be the effect of the Space Elevator on the rest of the economy?

In the next several sections we will try to answer that with a more detailed look at the types of things that the SE will make possible. The short answer is that space is cheap, it's only the first six or seven hundred billion dollars worth that is the hard part. We just need a certain critical mass of "presence" in that environment. With the ribbon we've got the breakthrough we need.

In general we can point to some effects. Its effect on the world, and especially the US, aerospace industry could be profound. Dead in their tracks would be the dreams of heavy lift boosters like the Shuttle C that we all had hoped for so fervently until now. But never fear, they are only delayed, we will need them later, not to lift off from Earth but to move ever larger masses from place to place in space. (Only a few heavy traffic destinations like Mars will have direct ribbon to ribbon transfers.) At the other end of the spectrum the stock of the small engine makers should soar. Now we will need small rockets by the thousands, and not just chemical rockets. The electric rocket motors of various kinds should be in great demand, quickly pushing their development to more useful thrust sizes in the 100 to 1000 Newton range. (They are slow but for cargo, oh, so economical.) And finally, if we can lose our prejudices, nuclear engines, the power and economy king of space travel.

By ordering nanotubes in thousand ton lots for the SE we will make them a commodity with a spreading effect on other products and industries beyond comprehension at this time. Just as one small example, everything we use cables for lifting will suddenly be 2 to 10 times more efficient. Or think of anything that uses fibers for strength with that kind of potential improvement.

From the advent of two, 20 ton ribbons, at the cost of less than $200/kg, space travel will be invigorated, to say the least. Today, satellites cost hundreds of millions, often more than the transportation to orbit which itself costs hundreds of millions. Their cost is driven by two factors. They have to be extremely rugged to withstand the shock of being launched; 3 to 5 g's acceleration and enough vibration that astronauts have their fillings checked after each flight. They also have to be operationally rugged, with redundancies and over engineering to make them last. Rare are the chances for repair, such as the Hubble. Without these requirements satellite costs could drop by 80% or more, and with just the 20 ton ribbon, launch costs could drop by 98%. With volume, now that space is more accessible, these costs would drop even more over the years, to the point, as we saw in the last chapter, of being an incidental expense. You still want your satellite to last, its just not as critical; replacements are readily at hand. And satellites can now afford to be built with stronger position keeping engines so that they can be maneuvered, to change position, to avoid collisions and to be de-orbited and destroyed in the atmosphere when they no longer work. We already have a space junk problem. (By the way, anything within 700km of Earth will slowly spiral down and burn up in 5 to 20 years if it isn't actively keeping its position, so if new satellites are built to de-orbit themselves we will eventually clean up the debris problem by attrition.)

And it only gets better from there. The 200 and 500 ton ribbons will not simply invigorate, they will totally transform space travel. Initially, cost would probably start out at one hundred dollars per kilogram and work its way down to the range of $3 to 10/kg, with a markup. In this range we can do almost any project in space, very economically and the transportation system is making a profit! That in itself is a radical concept. While we take a look at some of these activities, keep in mind another radical concept, for the first time we will be able to return materials and cargo to Earth, an idea hardly contemplated before.

LEO Traffic

You will recall that in chapter 7 on destinations we outlined the three classes of trips serviced by the ribbon. Let's review. We had trips to drop off satellites into LEO from altitudes less than GEO (36,000km) with two burn rocket insertions from below 24,000km and one burn insertion above that. We had trips to GEO and the return traffic. And we had trips beyond GEO to use the excess rotational velocity of the ribbon to sling ships to other destinations. At that time we were dealing with the 20 ton capacity ribbon. Here we will update these details in light of the availability of the larger ribbons.

Of the 13 ton payload (20 ton climbers) we could put into LEO a variety of satellite sizes, from: 8 tons of satellite, dropped at 30,000km down to only 1.4 tons of satellite with a short-shot maneuver at 500km. Converting to percentage terms so that we can work with any size ribbon we have an approximation table:

Table 14.1

Drop Altitude	% PL to 300km LEO
400km	10 %
14,000km	38 %
20,000km	48 %
30,000km	66%

Almost all the satellites now in LEO are in the range of 1 to 8 tons, all of which could be delivered to LEO off of the 20 ton ribbon. And the immediate effect of the SE prices dropping below $300/kg would be to render this crop of satellites, some 3000 of them, obsolete. They can write them off, take the tax deduction that will more than cover the 10%, or so, of what they would have had to eventually pay to replace them anyway, and send up new ones. Replace them all with much better quality, greater capacity, more features and extended staying power with more station keeping fuel. Plus, they can clean up after themselves by being equipped to de-orbit when no longer useful. This alone is enough traffic to keep the SE's busy for several years, but that is not all. Just for the telecommunications industry alone there is a reported desire for over 2000 satellites. The desire, not necessarily the capital. The collapse of Motorola's Iridium project has sobered these projections but, with much lower costs, this traffic would come back with a vengeance with the SE, producing several more years of full use traffic.

With the 200 ton ribbon we can deliver just about anything that might be wanted in LEO. Just using the regular climbers, 66, 50, 40 ton we can put significant tonnage into LEO. The max FLP, 66 ton capacity with payload of 46 tons, could put up to 30 tons into LEO from 30,000km, 20% more than the current Shuttle capacity. A full capacity 140 ton payload climber could insert 92 tons into LEO. Within this range (30 to 92 tons) we can send up modules to build any size LEO space station we like, for millions of dollars instead of billions, because the predominate cost is now the cost of the hardware. The short-shot trips work even better on the 200 ton ribbon. The 50 ton climber can scamper up to 400km in two hours, fire off a 4.3 ton satellite using 28 tons of fuel, and be back for lunch. These days, as technology makes things smaller, lighter and yet with more features, you can do a lot with 4.3, or even 1.4 tons of satellite, and the short-shot method makes emergency replacement of this size satellite possible.

By the time we add the next addition, a 500 ton ribbon, the max FLP climber would be a 166 ton climber, 116 ton payload, able to carry 70 to 100 people as a bus, or a combination manned/cargo carrier like the shuttle, only 60% larger. This is far better than trying to go to max loads on the 200 ton ribbon for the over 100 ton traffic needs. We would now be able to keep up the high density rate traffic on both the 200 and the 500 ton ribbons. It would then make more sense to build separate cargo only carriers and send up personnel in all pressurized habitat vehicles. In fact these could be independently operable sled type return vehicles. The sled type (like the shuttle) reentry vehicles could then return people to Earth the same way we do now, de-orbit hot reentry. That part of space travel is already fast and cheap, it gives us a choice of landing points and relieves down traffic on the ribbon.

Another feature of our development at this stage is the greatly increased overall safety factor. Several extra direct return vehicles (sled type) can be kept at orbiting satellites for emergency evacuation of stations and fast return to the surface, and also, help is just a short cycle away should support of various kinds be needed from the surface. Separating dangerous cargoes, like fuel, from manned vehicles is easy with separate ships for each and yet they can be delivered only hours apart for continuity of operations.

With low cost manned access to low orbits we can now afford to clean up the space junk that is already a hazard today and would be even more worthwhile to do, to avoid hazard to the ribbons. Piloted ships could go after the large pieces, and old satellites, give them a push, slowing them down enough to skim into the atmosphere and eventually reenter. It only takes a little speed difference to start a de-orbit. There is even a way to do this without expending rocket fuel (for pushing, not capturing); attach a long electrically conductive ribbon (nanotube ribbon?) to the de-orbit candidate and give the ribbon a little downward impetus so that it droops below the old satellite. It will set up a drag in the magnetic fields of Earth and slow down the satellite, (eventually). The small pieces can be hunted with lasers of decent size, operating from a nice power source, which we will discuss shortly. Give them a zap at the right angle, and they will slow down enough to spiral into the upper atmosphere, reenter and burn up. (If you zap them at the wrong angle, we are not sure what happens to them.)

With the 200 and 500 ton ribbons we now have excellent capacity for handling people traffic. But it is still expensive to put humans into space. What with all the life support equipment and supplies of food and water, each person equals 1000 to 2000kg (depending on where they are going and staying) of total mass that has to be transported and paid for. In addition, to put people into LEO from the SE now takes extra equipment, rockets and fuel, over just taking them up to GEO. So, after all this time of LEO traffic being much cheaper than GEO destinations, the situation is now reversed and manned activities will center on the higher GEO orbits as we will discuss in the next section.

Standing still in Style

[The view from GEO is one of standing still over the same spot on Earth while the Sun and Moon go around the sky.]

The first manned GEO space station that we built to service two way traffic on the first two ribbons (20 ton) would have been about the size of the International Space Station (ISS), (but far less expensive). Now we can build a Real Space Station(s) to service all the ribbons plus commuter traffic.

The ISS is built of cylindrical units that are limited in size by what the Shuttle can carry. Typical volume of a module is 137.4 cubic meters. But you get more internal usable space the larger the outside dimensions, so instead of 4.5 meters in diameter, our modules would be about 7.4m, and 10m long, the largest that is (currently) practical to air ship to our sea anchored ribbon. (We could do larger diameters if we sea shipped them to the ribbon but this example is big enough to make the point.) Then you weld two of these together, end to end, (at the ribbon site) for a really large, useful unit of 7.4 by 20 meters and 860 cubic meters of internal volume. Just one unit is 95% of the volume of the whole, multi-unit ISS! It is wide enough to divide into three decks, and one unit would be large enough to start a working station anywhere, not just for our operations. Then add a dozen or more units over the next year or two and you have some really great digs in the sky, enough space for a couple hundred people.

A 7.4x20m space station module would have 860 m^3 of interior space, and mass 66 tons shipping weight. It could house from 66 people in close quarters, such as crew, the way today's cruise ships are rigged, or 32 people, the way officers on large warships are accommodated. Under commercial conditions, allowing for common areas and facilities, you could count about 20 people per unit for planning purposes. A dozen such units would house up to 240 people and could be put up in less than two months.

In addition to space for ribbon crew and transferring climbers from the up ribbon to the down ribbon you have hotel space for the people coming and going off the ribbon, lots of outside mounting positions to rent for GEO satellites, and restaurants, laundry and SHOWERS for everybody. The station becomes a regular commercial center. Technicians for the GEO satellite renters can "walk" (or its equivalent in zero g) to the satellite mountings with replacement parts, or to up grade the equipment. How marvelous is that compared with current satellite replacement methods that could take a year from the time you know it needs replacing and $500 million in transportation costs?

In building space habitats you must use pressurized containers, and the easiest shapes to build, able to hold pressure, are the cylinder and the sphere. For some time it will be much easier to do the finishing, detail work on the ground and ship a ready to plug-in unit up the ribbon. This will limit our choice of shapes to cylinders and the size to what the climbers can carry as a whole unit. In the case of the 140 ton payload limit that would be a cylindrical station module of about 11x21m, 2000 cubic meters volume, a very spacious size unit to work with. For the tourist trade, something near this size would make it easier to reach the 80 to 90 cubic meters volume per person that is about the minimum that cruise ship passengers now expect (including common areas). The only thing holding back these large vessels would be the strength of materials to hold the pressure, and the cost; costs per unit volume would most likely increase with size.

There is another whole branch of habitat technology that needs to be explored, inflatables; flexible material that is inflated, perhaps with a rigid internal frame like a dirigible. NASA was going to include the first inflatable on the ISS and we heard that there were plans to spin it for artificial gravity. But, in scaling down the project (several times) it got canceled, so we have yet to gain any experience with this concept. The mass ratio makes these constructs intriguing; up to three times the volume per mass over hard walled construction, plus the advantage of fast assembly. The work done so far uses multiple layers of Kevlar material interleaved with spongy layers and metal foil, it's a giant "bullet proof vest" not just to hold the air in but to prevent puncture by most of the small meteoroids.

The reason for the layers of different material comes from the finding that if you piled on enough Kevlar layers you could stop anything. That turned out to be so many layers that you couldn't lift it, whereas, you don't need to make it so thick (and heavy) if you varied the density of the layers. (The meteor breaks up sooner, in its passage, dissipating its energy before it is able to break through.) With nanotube material becoming available we should be able to make a kind of "Super Kevlar" meteor armor for flexible walls and also to line the walls of rigid wall constructs. Once the fabrication techniques are perfected, the inflatables should dominate space construction. They would be cheaper, lighter, and safer than hard wall methods. And they are scalable, you can extend them to almost any size as your techniques improve, giving new meaning to our space hotels "ballroom".

With the meteor problem being well on its way to being solved, that leaves the other major space habitat problem, radiation. For short duration stays in space we have adequate protection, but if we are going to live and work there, we need to find new and better shielding, not only for people but also for the electronics. For projects like working on the Moon or Mars the answer is simple, just go underground. Three meters of dirt give the same protection as a sea level house on earth. For work in orbit, "underground" is not readily at hand, unless we want to haul around an awful lot of dirt, making our habitats more like mud huts. This will take some research work, we need a good shielding material that is lightweight. Right there we have a problem because the conventional way to stop hard radiation is to interpose dense mass, the more the better. So the best we are likely to get, early on, is material that is reasonably thin if not lightweight. The best intermediate solution seems to be selective protection rather than habitat protection, harden the housings for the electronic gear and make "storm shelters" at the center of the habitats equipped to hold all the personnel during high radiation Solar flares.

As soon as manned activity in space is advanced enough to handle detail work in orbit, we will be able to ship unfinished parts of units up the ribbon, and from then on size is no limit. The trouble is, doing detailed assembly work in orbit is hard to do. Turning a bolt in space is not a simple task, you turn the bolt and most likely you turn the other way and not the bolt. The bolt turner has to be anchored in some way, foot holds or body straps, such that the torque goes into the bolt. The hold-downs then become the problem, restricting movement and slowing down the work considerably. We need both new technology and new methods, these will take time to develop after we get up there and have a place to practice and test. For instance, we need light flexible space suits that don't cost the current $10 million each and also need to be custom made for each person. We need assembly and fetching robots that can scoot around on their own following scripts, and others that are remote controlled by "virtual" operators inside the station, plus new space welding equipment and heavy load (mass) handling equipment. (It's not so much the "lifting" that you need heavy duty equipment for, it's the stopping that's the problem.)

Once these abilities are achieved it will be a major milestone in space development, the dividing line between being a trail hiker who must soon return to civilization and a homesteader building his cabin. With the ribbon virtually nullifying the cost of transportation to space we can afford to pour thousands of man-hours of effort into the development and testing of these new tools. Now we can go to spheres of any size, ring and wheel shaped assembles that rotate for artificial gravity, anything. This will make possible the construction of processing, fabrication and manufacturing facilities in space. For instance, we could build steel plants and carbon fiber (including nanotube) structural members (girders), using the ores and ices returned from asteroids.

You can expand stations indefinitely because the controlling cost, like any other building, is the cost and economic utility, of the added units themselves, not the cost, as now, of getting them to orbit. They could build separate stations to operate other ribbon pairs, elsewhere, or to mount manned GEO satellite clusters for other positions around the globe. You just float them over to their new position. One could get carried away with this idea. Do you know Larry Niven's famous SF book,

"Ringworld"? Well we could eventually build a continuous GEO satellite and have a RingED World.

Sky Power

The largest industry likely to make use of the ribbon system and geo-sync stations as soon as they are available is the solar power satellite (SPS) industry. The idea is to make electrical energy from solar cells on a GEO platform orbiting in a fixed position. The electrical current is first converted to microwaves, beamed to a ground station, converted back into electrical current and fed into the electrical power grid. This is not a new idea, Peter Glaser wrote about the idea in 1968 then patented the concept in 1972. Then the DOE and NASA did a fairly comprehensive engineering study of this concept in the late 1970's, for which C. C. Kraft, laid out the basic structure.

The orbital unit would be 5km by 10km of solar cells that always faced the Sun. On one edge is mounted a 1x1km wave guide antenna pointing at a fixed spot on the earth to transmit the microwave energy.

This huge structure, 50 square kilometers by 0.5km thick for rigidity control, would mass 35,000 to 50,000 tons and there would be 60 of them to get the 300 GW's (then) needed for U.S. consumption, (more like 400 GW's now). From table 13.2 you can see that we could ship one such unit in 6 months if it were the only traffic. But no need to hurry that much, NASA figured 2 years for each station using 600 workers, so we would first have to build more living space. Good. More work for the ribbons!

And that ain't the half of it, as they say. The other half, on the ground, is even larger, in size, not tonnage. The rectifying antenna will physically cover an area of about 10km by 13km, but actually affect an area of 12km by 16km because of the exclusion area around the antenna that is not safe to live in. The incoming microwave radiation is concentrated in the center of the antenna at 280 watts per square meter, diminishing to the edges at 10 W/m^2, the minimum for effective conversion to electricity. But there is still some "over spray" and thus the need for the exclusion zone out to 12 by 16km, where the strength is down to 1 W/m^2. People can still use the exclusion zone, for farming, ranching, just not live there all the time. You can even venture inside the antenna area itself, people and animals can take up to 40 W/m^2 for short periods of time; its not hard radiation,* it's more like being close to a large oven. More recent thinking places the safe limit for the beam center much higher, up to 1000W/m^2, with a consensus settling at 500W/m^2. This would reduce the ground area needed for the antenna by half or more, greatly simplifying installations and site selection. The beam area could be much smaller, no more than a football field, but then you would definitely fry anything that came in contact; so it is a question of heat tolerance.   Sent to in-space users, a tight beam can be used.   *[Hard radiation is that which ionizes (high energy disruption), microwaves, no matter how strong, can't ionize.]

The antenna would be built up off the ground so that maintenance trucks can drive around underneath it. It would be made of an open mesh material so that sunlight and rain would still make it to the ground. In fact only about 2% of the microwave energy would pass through to the ground so a normal ecology should be able to be maintained, under the antenna. This is a definite improvement over the ground solar cell stations that would block sunlight over their entire area.   Also, this open weave construction is considerably less expensive than the ground solar panels. All together, with less area, less expense and less obstruction to simultaneous use, the antenna is much better than ground solar panels.

The NASA study (1978) rated the system as a 5 gigawatt (GW) power plant using a standard silicon cell efficiency of 15.7%, with overall efficiency to ground output of 7.8%. [Note: for comparison, large power plants are about 1GW, average is about half a GW.] At this power level NASA estimated a very wide range for the construction costs, in today's dollars, of $9,000 to $37,000 per kW or $45 to $185 billion for one unit. For that price, it also assumes that lift costs would have to get down in the range of $100 per kg. NASA soon abandoned the project as unreachable. In 1999 NASA had second thoughts and reopened the study.

Most of the parameters of that 1970's design can be improved upon; especially when you have 10 to 20 years before you can use it. The SE would certainly add to this impetus with 200 ton ribbon lift costs down in the $14/kg range. The biggest improvement can be had by using the new Gallium Arsenide (GaAs) solar cell materials, with up to 40% conversion to electricity.   That increases the output efficiency by 2.5 times, and with any improvement in transmission efficiency we could get up to 3 times better output. For the same size space platform we now have a 15 GW plant. A full days production would be 360 million kWh, worth $18 million at current prices ($0.05/kWh wholesale) and probably two to three times that (following general energy prices) by the time it is built. That's $542 million a month now, maybe $1.5 billion a month when built. That is enough to fund a $250 billion bond issue to build it. With that as the high range of what is workable, hopefully, it will cost a lot less.

With the SE able to put up cheap on-site housing, worker leave rotation, and materials transport, we should be able to lower the cost considerably. The cost of solar panel material, about ($3,000 per kW) $45 billion is by far the largest component. The other costs, transportation to GEO less than $1 billion, crew support $1 billion, crew salaries and transport only $133 million, hardly make a dent. That comes to $47 billion or $3,133 per kW, a debt that can be paid off in a little over 7 years with the revenue from that kilowatt. It's doable! And the SE has a long-term customer for over half its capacity. If this was demonstrated technology and the costs were firm, you could raise $100 billion for the first two units in one day on Wall Street

With 15 GW's per platform we would only need 20 such to meet the 300 GW capacity we want for the US, while new nuclear plants, if we are smart, fill the rest, (500 to 600 GW's needed by 2020) so that we can stop using hydrocarbon power stations. Then, of course, everyone will want one, so we will probably be in the power satellite constructing business for the rest of the century.

Being able to discontinue the use of fossil fuel power plants may well be the most important and valuable economic achievement of the 21st century, worth way more than whatever it costs to put up the ribbons to do it.

[Current power plant costs are about $550/kW for gas, $1500/kW for coal (now "clean" coal, they used to be a lot cheaper). This is the installation cost, then you have the fuel expense, coal is cheap which makes up for its higher cost of installation – therefore the final cost of daily output is close, with gas kilowatts only recently more expensive than coal.   The installation cost for the SPS is very high but its fuel cost is zero, still the SPS would not be competitive right now. Three things will probably happen to make SPS's workable. The cost of hydrocarbons will go up, reducing the difference with SPS energy. After the first half dozen, the installed cost of SPS will come down somewhat, and third, the cost of SE travel to orbit will fall even lower. Actually there is a fourth thing that we know will happen to make SPS's desirable, we will eventually run out of oil and gas.]

Extraterrestrial Power

Once established in space for our use here on Earth, SPS's will be become a very popular power source for everything else as well. Earth orbit SPS's can beam their power to space stations and other near Earth facilities. They would also be appropriate for Mars, but further out than that and the strength of sunlight becomes too weak for current technology to use effectively for large power loads. (At Jupiter, sun power is only 3.7% of near Earth.)   Future advances may extend their range of usefulness.

For use in space we can use tight beam microwave, or laser transmission, which makes the size of the receiver much easier to manage, a mere tens of meters, depending on distance. Our use of power beaming would push this narrow beam technology to the forefront making electric propulsion of spaceships operating in the near earth environment readily available. Without an onboard power supply the mass/propulsion ratio of such ships is quite "marvelous", removing yet another cost bottleneck to the development of space.

Our Moon, however, is a problem. Thanks to the Space Elevator, there are likely to be large facilities and lots of people on the Moon in this century and solar power is an obvious choice for power, the sunlight being just as strong on its surface as in space. And, land for solar fields is certainly no problem. The problem is the Sun sets for a 14 "day" night. This would mean a minimum of three solar panel fields spaced 120 degrees around the Moon to have at least one in full sunlight all the time. Then, for more than half the time you would have to send the power over land transmission lines one third of the way around the Moon (3640km) to maintain full power all the time. Transmitting power overland is a well-developed technology on Earth but we don't send it 3600km, unless it's an emergency, the line losses are prohibitive. Getting any greater efficiencies out of an old technology is not very promising, if they could, they would have already done it for the best of profits. So, solar power on the Moon is faced with low efficiencies and the redundancy of solar stations that can only work half the time.

SPS's would not appear to be much help. To be effective they have to stay over the same spot on the surface, that is, to be in synchronous orbit. Well wouldn't you know it, you would have to put the satellite way out here at the distance of the Earth from the Moon. Since the Moon presents the same face to the Earth all the time, we stay over the same spot on the Moon, we are in synchronous orbit with the Moon. The Moon rotates so slowly (27.3 days) that it doesn't have a synchronous orbit in the same sense as Earth. It has instead, five points (6, counting the Earth itself), the LaGrange, or L points, where objects are more or less stable and stationary with respect to the Moon. L5 and L4 are as far away as the Earth, and L3 is behind the Earth from the Moon. That leaves L1, between the Earth and Moon and L2 behind the Moon. L1 is also reasonably close to the Moon, 65,000km, a lot better than beaming power all the way from Earth. One small problem. If we put a satellite there, L1 is directly over the Moon's equator, it will pass through the Moons shadow once a month.

Never fear, the SE is here. The SE Company can fix this problem. We can't use a SE on the Moon for climber traffic, the Moon doesn't rotate fast enough (centripetal force too low) to support much weight on a ribbon. But we can use a very slim, light ribbon to tether an SPS just beyond L1. The reason we need a tether is that L1 is not quite stable, objects will slowly drift off the point and once off, start drifting faster. With the tether the station will tend to wander around the point. Actually we want it to do this; we may impart a little sideways kick to make the station go into what is called a "halo orbit" a slow circular rotation perpendicular to the ribbon. This will allow the station to see around the Moon and stay in the sunlight for those times that the Moon would block off the Sun (earth-side "night" is when we need it). This allows continuous sunlight on the satellite and continuous power beamed to the Moon's surface. We could do the same for installations on the backside of the Moon through L2.

Space Real-estate
Or how to make money from nothing.

The commercialization of space is very important – without it Man will never have a permanent presence in space. Along these lines, several companies, probably first among them the SE Company itself, will build space station facilities for rent or sale. There are three things to rent/sell in space. Nothing. That is, the vacuum itself, the real thing (take no substitutes), harder than any vacuum that can be produced in any lab on Earth. Zero gravity (or "micro-gravity"). And the view; the mounting of astronomical telescopes (of any wavelength!), and telecommunication (radio frequencies). (Detailed surveillance of Earth would be better done from LEO.) The old real-estate adage of "location, location, location" still applies.

Among the eager users of our spaces in space would be the research laboratories. They are in need of the hard vacuum, for which they are already paying a good deal of money back on Earth to try and get. Others will want the zero gravity conditions, for which no one is paying anything for back on Earth, and would be happy to have this as a new expense. Some will want both vacuum and zero g, we will call it the "absolutely nothing lab". The importance of these conditions is that the study of very small things is frustrated or distorted by the presence of gravity or of foreign molecules (that a vacuum would prevent). Scientists from the bio-tech industry need to study chemical reactions at the molecular level, the inner workings of cells and the hot subject lately, how proteins fold, (which determines their properties) without the distortion of gravity. The material scientists and physicists are studying such things as how crystals form, atom level activities and quantum cavities, in search of ever smaller and faster computer components. (Someday computers will be too small to see, no kidding.)

Such factories and lab's produce the highest value, lowest weight product possible, knowledge. Patents and techniques that will translate into new cash-flow products back on Earth. The rent would be higher than any Earthly lab, the working conditions are strange, and the commute is long, but the return per cubic meter of space rented could literally be in the tens of millions of dollars. New discoveries will undoubtedly result in new products on the molecular and atomic scale that can't be made in the Earth's environment. This leads to the first space industries.

Space Industries

We are not about to see coffee mugs stamped on the bottom with "Made in Space", or worse yet "Made Over Japan". Industry will most likely come about in a piecemeal fashion. First there will be parts supplies for the station and the tenant satellites, then work shops for onsite repairs, then for parts fabrication, "bending metal" as they say. But all this is in the service of the locals, to save time and transport costs, nothing for export. "Yes, and they will also brush their teeth, so what?" The fact is that all of these seemingly insignificant activities, including brushing one's teeth in less than five minutes, will be major accomplishments in space capabilities. To have, not astronauts, but normal people (technically trained of course) perform the normal daily activities is not at all easy. Living and working in space is going to be a little like an intelligent person with a brain injury, learning to walk and talk all over again. These "little" skills will be necessary before laboratories and factories can be established.

The first for-profit exports will have to be items of very high value per weight and things that can't be made in an Earth environment. There are generally two types of things that could meet these requirements, things that you make only a few of, and things that you make a lot of. That is, intermediate "tools" of production and mass production of end products. The first industry in space is most likely to be the first type, intermediate tools to produce something else. For instance, it might be very hard on Earth to get a process started, say a crystal of certain type, or to assemble, atom by atom, the mask (template) to mass produce computer circuits. You could make the first several units of these items and send them down to Earth to be put into the mass production process. Bio-technology molecular manipulation to make the first of an item or get a process started also fit in this category. Since so much depends on these first steps, these exports could be worth billions of dollars per kilogram. Meeting our high $/kg requirement is intuitively obvious.

Not so intuitively obvious is mass production in space. The good news is that the value barrier will have dropped considerably. Right now products would have to be worth $100,000 to $200,000 per kilogram to make space manufacturing work. With the SE this barrier can drop as low as \$20 to \$60 per kg where many more products would qualify. Still, it would have to be things that can't be made on Earth for that cost. Small valuable things are easy to think of, bio-tech products like molecules, cells, viruses and anti-viruses, proteins, and strands of DNA, also a variety of computer chip and electronic components the size of pin heads and smaller. The problem is that many of these items need both the unique conditions of space and the weightier conditions of Earth at different stages of their production. This will result in some rather complicated and very interesting facilities, a level of sophistication that will take years to develop, and will need a good deal more working room than research labs. This will come about after the large ribbons, into the second and third decades of the SE when adding another hundred people and 10,000 cubic meters of new facilities is just another business decision.

Beyond GEO

Most of our ribbon is out beyond GEO. This portion, two thirds of its length, makes the whole thing work by providing the centripetal, outward 'force' that holds up the lower portion below GEO. If the ribbon were cut at GEO the lower part would fall to Earth and the upper part would fly out away from Earth. By definition, any point on the ribbon beyond GEO is traveling faster, around the Earth, than needed to stay in an orbit at that distance. Just as dropping off the ribbon when you were below GEO brought you into a lower orbit around Earth, so dropping off out past GEO sends you out to a higher orbit around Earth. If you go out to the right distance on the ribbon and let go at the right time that higher orbit is going to exit the Earth's orbit and you are on your way to another planet. (For a more detailed discussion of ribbon launch altitudes and the orbital mechanics involved, refer back to chapter 7, Destinations, and additional material in the Addendum.)

With our space elevator facilities in place we can immediately start exploration of everything. We have recently become adept at miniaturizing components and sending proficient, yet lightweight probes, some less than one ton, to explore the solar system. One disadvantage to this sate-`light' approach is small power supplies and consequently slow data rate and limited content transmissions back to Earth. With the ribbon we can go wild, 5 and 10 ton robotic probes, could now carry everything, including the kitchen sink, to all the planets and moons in the solar system. For instance, we could send a heavy duty, TV camera-toting, core drilling, rover the size of a, well.. . a Rover, to Mars doing hundreds of times as much work but costing less to send than one of the small probes of today. We could afford to send one a month to different areas of Mars, and have them all meet, a year later, at Olympus Mons for a reunion.

We can make interplanetary ships much larger, more equipment and supplies, because it would not be limited by what the available second stage rockets would be able to boost into planetary transfer orbits. Now the ribbon does that. One of the most notable advantages of having higher mass allowances is in communication. We could afford to have larger, full spectrum, cameras. With more power we can transmit broadband return signals and even streaming real time video. This is extremely important to the scientific mission, an improvement in data gathering of 10, perhaps as much as 100 times over current capacities. This also saves money (even if we spend more on bigger probes) because the more robotic probes can do, the less is left for the really expensive manned missions to have to take up once they get going.

The recent concepts for mounting a Mars expedition are to make the return fuel there and to have that supply assured by being done two years before the manned trip gets there (Zubrin, "Mars Direct"). It's cheaper with the ribbon to just send a tanker loaded with the return fuel supply. This not only gives back, to payload use, many tons that would have been for seed fuel, but cuts off another chunk of development costs and gets rid of the two year delay waiting for the fuel maker to do its job before the manned trip. (We will still want to make fuel and lots of other things on Mars as part of developing Mars, but we can eliminate the startup delay and doing it roboticly at 100 millionkm.) The net effect on a program like "Mars Direct" is to double, even triple the amount of men and materials we can send at one time and to cut in half the time it takes to explore and establish the main Mars base. We would have the shipping capacity for space ship sizes from 46 tons on the 200 ton ribbon to 116 ton ships on the 500 ton ribbon if we stay within the FLP traffic size climbers; and up to 350 ton ships using max ribbon limit. We can ship an awful lot of people and equipment with these size ships. With just the 200 ton ribbons established, Mars could have a permanent base, dozens of people, and a zip code in three launch windows (6 years).

Doing the Mars Ribbon

In order to build a Space Elevator for Mars we must have an Earth ribbon with the capacity to lift at least a large fraction of a Martian ribbon in one piece or the ability to manufacture a Martian ribbon on orbit. Once we have this capability it will be feasible to build a Mars ribbon. When we build it will be a question of when we can get around to it given the concentration of time and money on the Earth system and also on how long it will take to do the preliminary exploration of Mars.

The launch window for economical trajectories to Mars comes around a little less than every two years. As soon as a Mars project is set in motion, a likely scenario would be to send several heavy duty robotic probes, as described above, for each of three such windows. It will take 6 months to get there, and at least 6 months to know how they are doing, and another 6 months to process the data into a comprehension of what they have found. If we sent three during the first window, say a week apart, we would have just enough time to evaluate their performance, before making changes and upgrades for the next three to be sent out in the coming window. Probes massing five to ten tons might cost as little as $50 to $100 million each, for an exploration budget between $500 million and $1 billion for 9 probes over three windows and 6 years, an extremely modest sum compared to today's exploration costs. Each probe should be able to last for several years and range over several hundred kilometers.

Now the question is, will that be enough to find out what we need to know? We need to know the best place to send the first manned exploration, which hopefully will become the main base on Mars. We would also like to know the next three or four best sites to explore on subsequent manned landings. In general we need to find the best places to obtain the mineral resources to develop Mars. We may find that it will take several scattered manned bases to get these resources, hopefully not too scattered. The most important criteria for determining the base site is finding water, or at least a good area to prospect for water. We know that there is water at the north polar ice cap, but the starting phases of developing Mars would be very difficult if we had to rely on that. Water may be in the form of subterranean ice or liquid in rock formations or, the best of all luck, hot water in geothermal formations, where we can also use it as a power source. In any case we will have to drill for it and that will take manned expeditions. We could luck out and find the perfect base camp with the first probes but it's best to plan on acquiring a large database of geological information over several years and many probes.

The first manned expedition could conceivably be done with two ships, one as return vehicle with backup supplies and support, the other with the people and the first set of supplies and equipment. One of their jobs will be to determine if the site is going to work as the main base or do we go for the next best site on the list. The next question they need to answer is where to put the Mars Elevator. Since the Elevator needs to be on the equator, (an alternate method will be discussed later) you can see that there could be a serious conflict between the best base camp and the Elevator site. Let's first deal with the aspects of the Mars ribbon itself.

Mars has 38% Earth gravity (0.38g) but the same rotation rate (24.6 hr.), the rotation we need for the centripetal force that makes the ribbon work. The lower gravity gives Mars a lower synchronous orbit, and a lower mass ratio for a ribbon of given lift capacity. About two thirds as long and about 1/3 the mass ratio for the same lift capacity. For instance, we can use an Earth 200 ton capacity ribbon to make, along side of it, a 53,000km ribbon massing 340 tons in only three passes using a full capacity climber. On Mars this ribbon would have a 71 ton climber capacity with a 50 ton payload. Now keep in mind we are talking mass here. A max load on Mars would be a 71 ton mass but only "weigh" 27 tons as far as the ribbon is concerned. This is about the right first-ribbon size for Mars; the Earth-Mars transfer ships are likely to be in the 30 to 50 ton range.

At 340 tons the first Mars ribbon is actually small compared to the thousands of tons of ribbon we have been handling. This gives us an alternate method of production. We could produce the whole thing at our ribbon making facility and ship it to the ribbon site in one piece. This would be much easier, cheaper and with better quality control than splicing with climbers. If we had the 500 ton ribbon we could just barely lift the whole ribbon to orbit. An alternative would be to unspool the ribbon, fold it at the thickest point and drag it up the ribbon, stringing it up alongside the working ribbon. Then we spool it up again, from the end to arrive at the GEO station in one large roll where we mate it up with its rocket engine and fuel tank assembly. This will take further studies however to determine feasibility.

For the plane change, maneuvering and circularizing into a GEO orbit at Mars it will take an engine and fuel platform of about 120 tons, which could be sent up the ribbon in one piece if needed. The whole package is now 460 tons but that is acceptable on ribbons of 200 tons or larger, beyond GEO where the maximum force on the ribbon never gets higher than about one tenth of a gravity. (And that is a negative one tenth, from centripetal force.) Now we run the package out to 54,000km on the Earth ribbon and sling it to Mars.

The next step is to bring up in parts the counterweight and its engine platform, assemble them at GEO and send them off to Mars, close on the heels of (or connected to) the ribbon spool. It will be about the same mass as the ribbon package just sent and would include supplies, fuel for ribbon deployment and maneuvering, climbers, and depending on design choices, could include the power station and laser beaming system. On Mars our options are to make the power for the laser beaming system with solar satellites or solar mirror concentrators at the counterweight. Since the power is "up there" it would be efficient to beam the power down to the climbers rather than up from the ground.

Once there we get it into Mars synchronous orbit, we join up the two shipments and deploy it the same way we did the first Earth ribbon. Now we have the option of anchoring the ribbon on flat terrain right next to a base camp on the equator, or anchor it to the top of one of the big volcanoes like Pavonis Mons, which is right on the equator. The ribbon can be of a simpler design than its Earth cousin, we don't have to guard against lightning or weather, except for dust storms. The reason for building on top of a volcano is that the tops of the taller ones are almost out of the atmosphere. The station would be above the dust storms and the fine grit that could be harmful to the ribbon itself, to the climbers and other equipment as well as blocking the laser power beam. Of course this assumes that we will have developed equipment that can easily handle zooming up and down the mountain, perhaps a railroad, and that we don't mind the time and labor, either. Local access on the flat, right next to a major base camp is definitely more convenient; it will be a contest between that and the cost of dust storms on the equipment.

By far the biggest problem will be with the two little moons, Phobos and Deimos. They are very low, 5,955km and 20,105 respectively and they orbit very near to the equatorial plane, which means continual near misses (or is it near hits) with the ribbon.

Some have suggested, (was it originally Clarke?), and more recently Robert Forward, that a wave action be set in motion up the ribbon to dodge the passing moons. It is true, that once you get the oscillation going it will continue for some time, only needing occasional energy input as it dampens out over time. But just how would you induce the wave in the first place, side firing rockets on a climber or drag the lower end up and down the plains of Mars? Well, we do have computers to figure just when to do this, but it still sounds a bit dicey. This will need more study. Another approach is to make the anchor point mobile as we did on earth only in this case it would have to be a road bed or even a railroad track of about 20 to 30km, built around the rim of the volcano, (it's large). The anchor point can then be moved as needed. The drawback to this plan is the rather large establishment of men, equipment and industry that would have to already be in place to build the movement system before you could have an elevator. That's the kind of construction job you would like to have the help of a ribbon already there. Keep in mind that with aerobraking in the atmosphere it is cheap enough to land ships on Mars, it's sending them back that is the expensive part, which is why you want an elevator. Otherwise you are stockpiling valuable space hardware on Mars.

Other suggestions are to eliminate the moons as missiles-in-prospect by moving one of them to the ribbon's Geo-sync point and make it a station or move them out to the end as a counterweight. The Geo-sync altitude for Mars is 17,040km, so Deimos, at 20,105km, and the smallest is the obvious candidate. But Deimos is not the main problem. With a period of 30.2 hours it catches up with the ribbon every 4.4 days and only a few of those crossings a year would be trouble. Whereas, Phobos is trouble all the time. With a period of 7.6 hours it does 3.2 ribbon pases a day and since it orbits near the equatorial plane, all of them are close calls. [Phobos is the answer to the trick question "when is a month shorter than a day?"]

Deimos is about 1.8 trillion tons and Phobos is about 10.8 trillion tons, mass. So if you want to start moving things around it will take a lot of rocket power. To move Deimos, the smallest and closest, down to Geo-sync orbit is only a ΔV of 0.223km/s. Looks easy. If you used liquid H_2/O_2 engines, Isp 450, with 7.5 million lbs. of thrust (the Saturn V) it would take 100 billion tons of fuel and 429 years to burn it. If you had a cluster of nuclear engines, Isp 1000, same total thrust, it would only take 45 billion tons of fuel but still 429 years. Assuming the same thrust; the higher Isp only uses less fuel. It would take a massive increase in thrust to shorten the time. Oh, and don't forget, you have to transport all that fuel up to the moon. You want the figures for moving the much heavier Phobos? Just multiply by the ratio 10.8/1.8. Anyone for an ORION project test case? It's clear we aren't going to be doing much moon moving this century. Another, "managing the problem" type idea, as a multi year project we could try increasing the inclination of Phobos's orbit so it comes at the ribbon at a wider angle, easier to manage moving the target when needed.

There is the possibility that we could get away with leaving the ribbons on the equator and yet not get hit by a moon. Recall that we have said that anything in orbit will eventually cross any given point on the equatorial plane, namely our ribbon. This is generally true for the crowded skies of Earth, but this is not true for all possible cases. An object could be in an orbit that had a harmonic relationship to the planets period of rotation. That would mean that the satellite would cross over some points on the equator again and again and not cross over others at all! One or more safety zones! At first blush there does not appear to be a harmonic, the orbit times don't divide evenly into each other, but a careful mapping of their orbits may turn up one.

Following the old English custom of "If you are going to be troublesome, I am going to leave", why not move the ribbon and avoid the moons entirely?

Let's say we move the anchor point to the north of the equator. (We have good reasons to prefer being in the Northern Hemisphere anyway.) This will shift the entire ribbon to that side of the equator, just like moving a chandelier, over, to hang from a different part of the ceiling. The ribbon counterweight will tend to "reach for" (to tangentially approach) the equatorial plane and the rest of the ribbon would trail down at an angle off the vertical to the anchor point. The way to see how this works is to work back from extreme conditions. The limiting case would be with an infinite rotation speed (of planet) or zero gravity. The ribbon would then stick straight out perpendicular to the planet's rotation axis independent of where on the surface the ribbon is anchored. In other words, as you move the anchor point, on such a high speed ball, the ribbon will stay parallel to the equatorial plane while the angle to the ground at the anchor point will change as you go toward the pole. (Parallel to the ground at the pole.) Now, working down from those conditions to something more like reality, as you lower the rotation speed and up the gravity, an off the equator ribbon begins to "drape" toward the equatorial plane.

Now, taking up our moon avoidance problem, the theoretical lower limit (without drape) of how far you would have to move the anchor would be the actual distance that the moons wander north and south of the equator in their orbits. To this we must add enough distance to allow for the drape of the ribbon back towards the equatorial plane. The moons orbits are off of the equator plane (0 degrees) by, Deimos 1.79 degrees and Phobos 1.1 degrees. By moving the ribbon anchor point north of the equator, you would allow the moons to pass on the same side, south, of the ribbon all the time, there would be no interception point and no further need to move the anchor station. Deimos with the largest displacement, 1.79 degrees at 20,105km, this is a 734km wandering north and south of the equatorial plane. Phobos swings 180km from the equator. If we can miss Deimos we don't have to worry about Phobos. Now the question becomes, how much will the system tend to lean or bow toward the equator? As a first cut at the calculations it looks like the ribbon sag is only about 3km. If we plan on adding 20km to the 734km we should be in good shape.

From the elements involved in the calculations it seems that the reason for this small displacement of only 3km, is that the mass of the ribbon over the tension in the ribbon is what causes the drape and this ratio isn't much. And once you get out beyond a few Mars radii the gravity vector (vertical) is now very small and nearly parallel to the equatorial plane so the counterweight and most of the ribbon won't be pulling sideways. The bottom line is that we can pretty much use the limiting case of high rotation and small gravity as a close approximation. This also means that the added (offline) stress is almost nothing. The only other question is how much a climber will drag the ribbon down (over) toward the equatorial plane. If it sat there at about 1000km it would add more of a drape but if we move it quickly it should only be a momentary movement. Long before we go to Mars we will have good experience with the forces involved with dragging a ribbon around and will be able to easily add what little extra margin the Mars ribbon might need. It is interesting to note that Olympus Mons extends from about 704km to 1190km north of the equator.

One of the disadvantages of all the above methods is that we need a large presence on Mars and in some cases a good deal of construction before we can partake of the fruits of the elevator. There might be a way to have the ribbon first or at least early on in human occupied Mars. Suppose that we arrange for the ribbon deployment assembly to enter an orbit tilted to the equatorial plane a little more than the moons, say 2 degrees, and higher than Deimos. This is above synchronous orbit, but as the ribbon is deployed its orbital altitude will drop. On the bottom end we will have a good size mass with splayed legs that can shoot anchors into the surface to fix the ribbon. As the ribbon end gets near the surface we wait until the orbit swings out away from the equator, making sure that we are 754km or more north, and then set the end down and shoot the anchors. The ribbon at this point will oscillate across the equatorial plane and back, following the orbital path of the counterweight. A short propellant firing at the counterweight or possibly from a climber will eliminate the oscillations. We now have a ribbon standing off the equatorial plane, out of the way of the moons and ready to go with minimum human support from the ground, other than finding the right spot to put it, near a good base camp site.

There is another arrangement we might try, for the future when we would have many ribbons on Mars and we want to simplify things as much as possible. We could spend the time and energy to straighten out the moons orbits so that they were right over the equator, then we need only move the ribbons off the equator by just a little bit, lessening the problems of odd angle forces. For those of you that like to work with large numbers the ΔV's for moving the moons into an equatorial plane are Deimos, 0.042285km/s, and Phobos, 0.020467km/s. Ok, here is one answer. Using, as we did before, as a standard unit of thrust, one Saturn V rocket, namely 7.5 million lbs. of thrust, it will take: Deimos, 72.7 years and 17 billion tons of fuel, Phobos, 210.5 years and 50 billion tons of fuel.

As we said, this is for the future. But keep in mind that if politicians do not stop Man's progress the energy at our disposal will continue to grow and by this time next century we would have the energy budgets to do a little moon moving on the weekends. The two easiest moves would be flattening the moons orbital inclination, where even part of a move could be useful, and altitude (velocity) moves that would change the period, with the possibility of setting up harmonic spacing, creating safety zones.

Mars traffic might seem a little slow with just one ribbon but it could easily handle one 50 ton (mass) shipment every few days, and this would imply a large crew to off-load and disperse the incoming goods and another large crew to use the goods to build, whatever. In addition we would have the outgoing traffic, and maintenance of climbers, Mars would be a busy, fast growing operation to handle all this. The 50 tons is a nice size ship for Mars work. It would allow the movement of 8 to 10 people at a time and almost any of the heavier pieces of equipment they might want in one piece.

Soon Mars will need a second ribbon, then a heavier one, colonization will be well under way and the traffic with Earth will be an every day occurrence.

Chapter 15:
Space Tourism

"Now Johnny! Stop fiddling with Grandma's jetpack. You're going to turn her into a shooting star."

"No, we are not <u>there</u> yet! We are only just past the Moon."

Ah, the travails of family travel.

Will people ever take vacations in space? It seems that there are plenty of people that want to. In 1993 a survey was done in Japan on the public's interest in personal space travel. Surprisingly, most people wanted to go. Wondering if this was just a fluke, more science fiction fans or perhaps a desire by a lot of Japanese to get off a crowded island, a team headed by Patrick Collins, Masataka Maita and Richard Stockmans decided to find out what the North Americans thought. So, in 1995 they conducted a survey of the US and Canada. Here are the key parts of their report. (Edited for brevity. Text in brackets, [ ] are paraphrasing of longer passages or our comments.) Keep in mind that this report was written in the context of 1995, when the only known method of getting into space was with chemical fueled rockets.

Demand for Space Tourism in America and Japan
And its Implications for Future Space Activities

Introduction

[If space tourism is to be feasible there must be sufficient demand. What is the overall interest and how much would people be willing to pay for these services?]

In 1993, the first market research on the demand for space tourism was carried out in Japan, supported by the National Aerospace Laboratory (NAL). That survey of 3030 Japanese people across all age groups revealed a surprisingly strong popular wish to visit space. More than 70% of those under 60 years old, and more than 80% of those under 40 years old, stated that they would like to visit space at least once in their lifetime. Furthermore, some

70% of these said that they were prepared to pay up to three months' salary for such a trip.

Although estimates have been published before, this is the first actual market research of its type to be conducted in America. The results illustrate that interest in traveling to space is also very high with North Americans. Overall, 60% of the people surveyed were interested in traveling to space for a vacation.
These results have significant implications for the future of space activities; in that the resulting estimates of the potential size of the civilian space travel market far exceed estimates for any other commercial space market.[1] As a consequence, many of the current vehicle designs for future space transportation systems may need significant design adjustments. [Just what did you have in mind?]

Age and gender. As an overall average, more than 60% of the population say they are interested, with a higher proportion of younger people being interested, as one might expect. The overall average figure is comprised of more than 75% of those under 40 years old; 60% of those between 40 and 60; and more than 25% of those between 60 and 80. It is notable that in every age group men were more interested than women, the average difference being about 10%. This is rather different from the results of the survey in Japan, where no significant difference was found. This may be due to the fact that many US astronauts are military staff, whereas all the Japanese who have visited space are civilians. [This is a rather dubious connection, world wide, most pilots are civilian males, so what?]

Length of stay. ["For the discount rate you must stay over for one lunar eclipse."] A substantial majority of all age-groups say that they would prefer to go to space for either several days, one week or more. In all age-groups, no more than about 15% of the people say they would prefer a visit of a few hours or one day; the majority have a clear preference for a longer stay, some 30% of those under 50 preferring "several weeks". The questionnaire did not discuss the relative price of different lengths of visit. In reality, a longer stay will be more expensive, but these results [alone] give no information on participants' price-sensitivity. It seems reasonable to conclude that, in the absence of orbital accommodation enabling people to stay for a few days, the demand for space tourism will not reach its full potential. [?? In the absence of "orbital accommodation" there won't be any space tourism, let alone full potential!]

Price Sensitivity. The first point to note is that there is little difference between the various age groups on this variable. At the upper end of the range, it is interesting that 2.7% of those wishing to visit space (representing almost 3 million people) say that they would pay three years' salary. Clearly for these people traveling to space is a very strong desire. 10.6% (representing 11 million people) say they would pay a year's salary, which is still a very substantial expense. 18.2% of participants (representing 19 million people) said they would pay six months' salary, and 45.6% (representing 48 million people) say they would pay three months' salary.

The overall average income for North Americans is approximately $2,000 per month, or $12,000 for six months' salary, and $24,000 for a year.[2] In addition, two thirds of those wishing to visit space said that they would like to do so several times, not once only. Consequently, even allowing for a substantial gap between consumers' intentions and their actions, space travel could clearly be a multi-billion dollar per year market in America alone. Ten percent of the above estimate is still more than $60 billion in funds that people say they are willing to spend on a space trip.

Reasons for not wishing to visit space. Roughly 1/3 of each age group said that they are simply not interested in the idea. It seems likely that these people would not go, even if the service were available. However, some 5 - 10% said that they consider the idea unrealistic, and presumably some of these would wish to go once the service was really available. Roughly 1/3 were concerned primarily about safety, which emphasizes the over-riding need for reliable space transportation. Some of these would presumably reconsider once the service was something like that of today's airline travel. A substantial fraction of these respondents however, said that they were afraid of flying, and so it must be assumed that they would remain unwilling under any circumstances. Finally, most of the people who considered themselves too old to go, were still interested in traveling to space if it had been available when they were young. Overall, it seems that the average of 60% interested in visiting space could grow to 70% or more in the event that a regular service became available.

A weakness in terms of complete consistency between all respondents is suspected by the authors in terms of personal interpretation of the space tourism concept statement. Since the respondent was free to envision what the space trip and vacation would entail, it was felt that in some cases the respondent thought they might be traveling to far away places or planets on the trip. Whenever questions arose as to where one might be traveling on a space trip, it was suggested by the interviewers that these trips would be mainly in orbit around the earth. Although some misinterpretation may be present, it is felt that most respondents had a clear picture as to what would be involved; a vacation orbiting the earth or slightly beyond. Discussions with the interviewers during and after the study attempted to elicit information concerning this issue. If bias is present due to the lack of clarity involved, it could move the results in either direction, but probably not by a substantial amount.

Implications
From surveys of the world's two largest consumer markets, and supported by data from Europe, we provisionally conclude that if the price of a return ticket to low Earth orbit could be reduced to between $10,000 and $20,000 per person, a substantial fraction of the populations of the rich countries, of the order of 100 million people, would wish to make such a trip.[3] In this case the scale of the resulting business would seem to be limited primarily by how fast the capacity could grow, but the world-wide demand for short visits to low Earth orbit over time could apparently reach the scale of one million passengers per year or more, generating annual revenues of $10 billion or more.

Based on this demand, what can we say about the economic feasibility of developing a transportation system for space tourism? A market of the order of 1 million passengers per year paying on the order of $10,000 per person would generate revenues of the order of $10 billion per year. Thus the market potential of space tourism is somewhat similar to that of the new generation SST vehicles. In both cases of course, there is a very delicate balance between the scale of the actual market and the affordable cost of development, production and operation. In addition, the uncertainty concerning the actual level of demand in different countries, and the potential for revenue increase through market segmentation is still high, and needs to be investigated more thoroughly.

In order to reach the low launch costs needed to create a space tourism business, the economies of scale possible from accessing the entire global market will also be very important. Thus, from a business point of view, a collaborative, international civil project has the best possibility of success through supplying the global market. Success will also depend on achieving safe and comfortable flight conditions, comparable to air travel today. Such low launch costs, are of course a challenging objective, and the subject of "space tourism" is still controversial. For example, in 1994 NASA's Office of Advanced Concepts and Technology described space tourism as "only a flight of fantasy". However, since then NASA has agreed to participate in a joint study of space tourism with the Space Transportation Association.

To date, very little research has been done on passenger vehicles suitable for commercial service.[4] Consequently, more work needs to be done to study how reusable launch vehicle systems can reach the necessary operating cost targets. Estimates of the development cost of reusable passenger launch vehicles are also uncertain, ranging from billions of $US to tens of billions of $US. Yet even the highest of these figures are less than the cost of the international space station project alone, while the demand for tourism apparently has the potential to grow to many times the most optimistic projections for existing commercial space activities.[5]

Image Change

Today government space agencies face declining budgets, with a clear prospect of further cuts, fundamentally due to lack of popular support in the new post-cold war era, for the work that these agencies do. In this situation, trying to do what the public say they want, even if it seems difficult, is surely a good direction to take. In public opinion polls only about 20% of Americans describe themselves as "very interested in space". But this recent market research shows that the large majority of people, 60% "want to visit space for themselves", and would be prepared to pay high prices for such a service. It therefore seems probable that many of the public would prefer public money to be spent trying to reach this target even if it is difficult, than on projects such as advanced satellite technology or a space station for government researchers.

A scenario of space tourism development provides an interesting alternative vision of future space activities, one in which passenger space travel might become the major space activity. Business competition between travel companies, similar to that between airlines in the early days of aviation, would progressively increase demand and reduce costs. The fleets of passenger launch vehicles and the orbital accommodation needed to satisfy the demand of hundreds of thousands of passengers per year would represent major new markets for manufacturing and operating companies, and would be the first commercial demand for crewed space activities. This would be a major and popular change of image from today's current space activities.

Authors:
Patrick Collins - Visiting Researcher at NAL and Tokyo University, RCAST
Masataka Maita - National Aerospace Laboratory (NAL)
Richard Stockmans - York University, Schulich School of Business, Toronto, Canada

----

As toilers in the space vineyards, we of all people are happy to promote and delighted to see a strong interest in space tourism. They are undoubtedly correct that a broad public interest is beneficial to the development of space. And we could sure use the traffic to bolster the SE's use rate. So it is with some reluctance that we critique this rosy picture.

First, let us make it clear that there will be space tourism, the only question is when. We have already had our first space tourist in 2001, round trip to the ISS, for a reported $20 million, so we know it can be done, its only a question of the authorities allowing it and a lot of legal work on liability waivers. Note that it was the pragmatic Russians, not the regulationist Americans that permitted the trip. But one, or even a trickle of several more, do not an industry make. These examples are simply hitchhikers on the marginal capacity available, no extra trips or vehicles for passengers are needed.

This report's upbeat view of launching oneself into space is not unique. In the years since the report, perhaps because of it, there have been several articles in the popular press about space travel for pleasure. At one time you could make reservations (with deposit?) for a ride in space and for the Hotel Hilton's proposed multi-storied Moon Hilton; it may still be available if you hurry. There is even NASA art depicting freestanding aboveground buildings with floor to ceiling windows showing the grand vista. More fiction than science. Such windows would provide more than enough heat and radiation to cook the inhabitants. And it would take extremely heavy, very expensive engineering with materials brought from Earth to build large, rectangular, multi-storied structures strong enough to keep the internal air pressure from blowing them apart; very impractical. One article blithely suggested that the travelers would stop off briefly at the Moon for refueling on their way to Mars. This, presumably, from the same planners that route airline trips from Dallas to New York through Denver. (Delta V LEO to Mars, 4.5km/s, LEO to the Moon, 6.0km/s, so you are already behind when you get on the Moon and you still have to lift off and go to Mars.)

The most interesting finding of the report was that only 5 - 10% got it right, they said that they consider the idea unrealistic. Quite. The main problem with this survey is that there was so little context for the questions. It mainly showed that people would like to go on a vacation trip. If you had asked if people would like to go on something grand, like an around the world cruise, you would probably have gotten the same response. They did not ask the people what their doctors and insurance companies would think of them enduring three gravities of acceleration. (At the start of air travel it was not uncommon for those contemplating trips to seek the advice of their doctor. Flying was rigorous, high altitude without pressurization, turbulence, and sharp changes in temperature, which is perhaps why the first flight attendants were nurses.)

They did not ask them how they would like being airsick (space sick?) for a couple of weeks at a time, which over half of the population would suffer.   (Pilots, hence astronauts, suffer less but still a high percentage.) They did not ask them how they would like being confined in a container with no more than 5 to 10 meters of movement, in any one direction, none of it walking, for several weeks, with a couple dozen other people that do not shower as they do at home. They did not ask if the people would like to take a vacation, without insurance, in an experimental craft (a spacecraft). That is the rating a craft gets without a FAA airworthiness certificate, which a rocket plane could never get without a great deal of new law. (By the time you go down the list of FAA safety requirements that a rocket/plane couldn't pass, from landing gear tests to systems redundancy to emergency exits to landing go-around, its staggering; it would be easier to have no law, complete exemption. Bureaucracies are not noted for capitulation.) This is the context of the question about tourism as long as we have rocket craft taking off from Earth.

There are other errors of perception in the report.

[1] "...the potential size of the civilian space travel market far exceed estimates for any other commercial space market."

This is an unfair comparison, it compares a count of whims-of-the-future amongst the general population as against a list of realistic doable projects from a few business planners. It could not turn out to be true, that is, it could not be that more will be spent on tourism than on commercial development. By the laws of the way things work massive amounts of commercial development will have to occur before a surplus, the residuals, of time, energy and money is available to be devoted to tourism.

[2] "The overall average income for North Americans is approximately $2,000 per month, ... $24,000 for a year."

That is true, it is the average, but you only get to count, for vacation travel prospects, those pay scales that could afford it, a much smaller number for your "market".

[3] "From surveys of the world's two largest consumer markets, and supported by data from Europe, we provisionally conclude that if the price of a return ticket to low Earth orbit could be reduced to between $10,000 and $20,000 per person, a substantial fraction of the populations of the rich countries, of the order of 100 million people, would wish to make such a trip."

That's nice. That's $1 to $2 trillion, 15 to 30 years worth of US airline revenue traffic; some other consumer industry is going to have a sharp decline. That's more accommodations than all the hotel rooms in the whole world. At a rough guess it would take an entire year of the worlds GDP to build the facilities for these people before we could enjoy this revenue.

[4] "To date, very little research has been done on passenger vehicles suitable for commercial service."

True. This is an engineering non-sequitur. As just mentioned, it is against the law to carry paying passengers on any rocket ship, so who would bother to do research on it? On the other hand there is lots of research being done on all manner of space technology, some of it is even people carrying technology, for workers, so we can build things in space. Could there be more of it? Sure, space development is going far too slowly, we need to get on with it in a much bigger way, but it will not be for tourists; see next item.

[5] "Consequently, more work needs to be done to study how reusable launch vehicle systems can reach the necessary operating cost targets. Estimates of the development cost of reusable passenger launch vehicles are also uncertain, ranging from billions of $US to tens of billions of $US. Yet even the highest of these figures are less than the cost of the international space station project alone, while the demand for tourism apparently has the potential to grow to many times the most optimistic projections for existing commercial space activities."

No study is necessary; engineers will tell you straightaway that chemical rockets can never reach those cost targets. It is not a matter of a better design it's a matter of physics. And as to the costs? The new AirBus, 560 passenger, A380 will take up to $11 billion to develop and that is just an airplane, old technology. We don't know their planned margins but it should take sales of 3 to 4 hundred of them just to reach breakeven, not counting interest on that money while waiting for the sales to happen. Developing a passenger carrying rocket plane would be in the hundreds of billions of dollars, not tens of billions. You would have to sell many thousand of them to reach breakeven, more than the total number of airplanes in the worlds commercial air fleet. And then, since they can't be used at conventional airports, there is the problem of where to put them all. All that for only a $10 billion a year business? That is equivalent to the low end of middle size companies on the stock

market. Florida alone has about $50 billion a year in tourist trade.
It would never pay.

It is important that people understand the sequence of human
progress, tourism can't come before, or exceed commercial development.
It would be like developing hotels before homes and apartments and
having more hotel space than dwellings. The report writers do not seem
to understand the tourist industry. Touring is a subsection of leisure
travel, which is anything that is not business travel. It develops in stages.
Whatever the means of travel, like ships, the first uses are for urgent or
serious travel, exploring, colonizing, shipping cargo, pirating. Only after
people are well established in their new locations, and have surplus
income, and the industry has grown large enough for per passenger costs
to drop, does leisure travel become a significant percentage of that travel
method. The first leisure travel is mostly personal or family business,
going to visit people not places. Only after further increases in personal
income, and another drop in per passenger costs does true tourism, going
to see places or just going, come on strong. Tourism might be defined as
going places where you don't know anyone.   For the airlines, leisure
travel became the dominant traffic sometime in the early 1960's (over
business travel) but tourism, a subset of leisure travel is still not larger
than the casual travel to see family and friends.

Therefore, space accommodations must be built by business
travel, (freight and workers) then it must become occupied with residence,
(colonization, still business travel) before leisure travel can set in, finally
followed by tourism. One clue that the colonization phase is substantially
over is when one-way tickets die down; for space that may be a long time.

From the point of view of 1995 and rocket craft, the report is off
target. Having said all that, it is now ironic that from the current point of
view, with the prospects of a Space Elevator, the report may well be back
in focus on a serious market for space traffic. The SE has a good chance
of providing a substantial increase in commercial and scientific traffic, a
high rate of worker traffic and a high rate of building accommodations for
all those people. (Worker traffic can go on vehicles that would not be
FAA approved for public traffic.) Secondly, the climber vehicles, being
without volatile fuels, and not being "air" craft in the strict sense, have a
realistic chance of being able to be built to public transport quality and
safety standards for carrying people for hire. In other words, the SE is the
only hope of meeting all the preliminary development requirements for
having tourism in this century.

Traffic

There won't be any room for visitors until we get the large orbital stations built which will be after the pair of 200 ton ribbons are built. According to our schedule this would put us in the $14/kg price range. Note that this is less than one percent, actually 0.14% of today's costs. Yet the cost per person would still be $33,600 round trip. (That is counting a person as 1200kg of total load to be transported, their share of the total vehicle and life support equipment lifted. And that is very generous, 2000kg per person is probably wiser accounting.) A few visitors might come, VIP's on special trips, way short of regular passenger service.

How will it be once we get the pair of 500 ton ribbons up? That will put the cost in the $9/kg range and a total yearly capacity of 289,000 tons. People round trips would then be $21,600, lots better but still a pretty select market. (Don't forget, these numbers do not included food and beverage service, plug in a fair cost for two weeks of your eating style.) At this stage of development we would be in the midst of large building projects, solar power satellites for one, and the on-orbit living, support facilities, and worker traffic so involved. This could easily use up the entire SE yearly capacity. If there were a couple of thousand tons capacity left over we could carry a couple of thousand passengers. That would be about the number of people in the world that could come up with $22 thousand, plus bread and water, for the trip.

Realistically, it doesn't look like full service passenger traffic is going to be available until we are well into the second pair of 500 ton ribbons. This would be some 20 years or more into life with the SE. The first, high cost, ribbons will have been paid off and our construction overhead, current operating costs plus markup will be in the range of $10/kg retail, or $20,000 per person, just within the range the tourist promoters are looking for. This would actually be closer to $30,000 counting food and hotel services for a week or two in orbit. At this point, with 6 large ribbons operating, our yearly capacity would be 494,000 tons. Commercial and construction projects would have driven the SE system to this level. If even one quarter of capacity were available for passenger traffic we would be surprised. That would be 110,000 passengers a year, about $3 billion in sales, a modest start to leisure space travel. (Note that the cruise ship industry handles about that many people every two weeks.) The lack of a more robust tourist traffic is not that we don't like the tourists dollars, on the contrary, we would make more margin on that traffic, it's a question of who came first. The ribbon system will expand by selling the expectations of future capacity at future, lower prices. Contractors will commit to this capacity years in advance; people don't buy vacations that way, so commercial traffic will dominate. In addition, there has to be surplus cash, capacity and time to make it worthwhile to build the more expensive passenger carrying vehicles. This must come from robust commercial traffic.

Traffic of the kind envisioned by the tourist promoters might be possible in the third or fourth decade of ribbon operations as lift rates per kilogram approach $3, one-week trips would approach $10,000. The ribbon system will have grown to a capacity in the one million tons a year range and we might even cross the line and have more than half of that, 500,000 people, $5 billion a year, as passenger traffic. By this time there would be at least worker traffic, perhaps passenger traffic with the Moon and Mars. Now you have the start of a decent tourist industry.

No rocket/plane is ever going to approach the $14/kg price we started this example with, let alone $3/kg. It will take the Space Elevator to have a space tourist industry.

Up the Beanstalk

What would space travel be like? From TV news and films, most people are familiar with the image of the rocket ride to LEO. That would be exciting, if not heart stopping for the physically unprepared. And that is what people now have in mind, the greatest roller coaster ride imaginable, when they think of space travel. By comparison a trip on the Space Elevator would come off as rather dull. But of course that is the point, it can be made safer and therefore accessible by the public.

The SE would be an elevator ride, a very long one, 7.5 days to get to GEO. The view would be spectacular. The first part of the trip would seem very familiar to air travelers, as you rose quickly up to cloud level and then beyond. As you passed through 12km the curvature of the Earth will start to be noticeable on the horizon. As you pass through 50km you will be out of most of the atmosphere and the sky will turn black, stars will come out, even in daylight, on the side away from the Sun. Now the most valuable aspect of the trip, the view, becomes a liability.

Actually it is the window that is the problem. Remember how small the windows are in jet airplanes? This is to maintain wall strength against the pressurization that keeps us comfortable at 35 thousand feet. The first pressurized planes had large windows; the view was an important sales item. Then it was found that the flexing of the hull with the pressure changes from flight to ground plus the large opening in the hull for the window added up to fuselages that tended to come apart in the air; bad for business. So now we get safer but skimpy windows. (You will also notice, if you look closely, that they have three panes, they are strong little windows.) For our spaceship/climber this is going to be a problem, the pressure difference is now three times as high yet we will want larger windows not smaller ones. The view is about all there is of value on this long trip, we will have to have large windows, even bay style windows, that people can sit around in groups. Then there is the second window problem, sunlight. Once above the atmosphere, we can't let direct sunlight in the windows and we can't leave it to passengers to close the shades or whatever. There must be an automatic way to block or filter out direct sunlight. Well, the technology boys and girls have some work to do and we will leave it to them; just so you know that passenger traffic isn't going to be so easy.

One solution to the window problem is not to have any. Although we don't think it would be as aesthetically pleasing for the first part of the trip, they could have TV "windows" instead. With the advances in TV technology already at hand and bound to be even better in the future we could have large TV "windows" on the walls of the passengers cabins. With something like 6000 by 4000 pixel detail that would be as good as, if not better than movie theater fidelity. With long lenses for magnification they would be better than regular windows for picking out details on the ground and of the stars. There could be a variety of cameras with different views with the passengers selecting which channel they wish to watch. The stars would be quite spectacular. Without an atmosphere and with an absolutely black background the stars would stand out crystal clear, and twice as dense as the best seeing conditions on earth; we will be able to see all the way down to an eighth magnitude, before camera magnification! (Since most people haven't lived in "the best seeing conditions" the increase in the multitude of heavenly bodies would be really spectacular.) And we would notice that they come in colors. Some of the cameras could be "for rent" where a passenger can use it for a while and drive it around with a joystick, looking at whatever they want. We will have to be set up to receive all the commercial TV channels anyway to kill time on such a long journey.

At 100km, if you are a sensitive person, and are able to move around a bit, you will begin to notice the lower gravity. At 346km it will be down by 10% and most people should notice. At 753km 20% lower, at 2643 50% lower and at 14,000km 90% lower. There will probably be wall mounted instruments showing altitude and gravity to keep the passengers informed, and games will develop for the amusement of displaying the change in gravity. At least we have gravity, this is a great advantage over the rocket trip and the difficulty over half the population would have with weightlessness. We can't test it before we actually build a ribbon, but the slow reduction of gravity during a multi-day trip may well serve to significantly reduce the number of people adversely effected by zero gravity by the time they get to GEO station.

While the SE keeps a firm footing under our passengers for most of the trip and delays the problem of zero gravity until close to GEO station, another problem of space can't be avoided, pressurization. The ideal situation would be to just have standard sea level pressure all the time, but that is probably going to be too expensive. To maintain standard pressure would force the use of much stronger and heavier materials and thus more expensive construction. High pressure also increases the safety tolerance problems prompting another round of heavier construction - it all compounds. The airlines pressurize to 5,000 ft as a compromise to this strength to weight trade off, and usually the public doesn't notice. Until, that is, the pumps let it slip up to 6,000 feet and because of traffic the plane descends rapidly from cruise to approach altitude, then everybody notices and you can hear the grumbling as passengers file down the exit ramps tugging on their earlobes. We don't need sea level pressure, as witness people that live in the mountains. NASA uses 5 pounds of pressure, one third of normal, equivalent to 27,000 ft altitude, just short of the top of Mt. Everest. But then you have to up the oxygen mixture to 60%, three times normal, to maintain vapor pressure on the lungs so that they absorb oxygen at the normal rate. This seems to be a comfortable trade off for healthy people, and a huge load off the engineering budget for pressurized vessels. But not all lungs are the same. What percentage of the public would have a problem with this? Would we have to throw people into a pressure chamber and test them first? Bring a note from your doctor?

While you practice your breathing, the next problem subject is food, or rather the cooking of food. It has long been a tradition of high priced travel that meals be luxurious, especially the serving of food "to die for". At low air pressure water doesn't boil at 100 C degrees. At NASA standard pressure, 5lbs, or about 345 millibars, it boils at 72 C degrees, too low to break down the cells and proteins of food, so that they do not cook, they just get warm. Pressure cookers would be necessary, which might drastically limit culinary skills. Ambiance might also suffer, in 60% oxygen there would be no candle lit dinners or after dinner cigars and brandy. Oils are not compatible with 60% oxygen so most of the personal grooming products, especially perfumes are out, or would have to be reinvented. Even clothing would be a problem, certain fibers would deteriorate or even spontaneously combust in 60% oxygen, so guests would have to be issued all special clothing for the trip. Hardly the fashion statement most travelers would look forward to.

Once we are above 150km it will become very noticeable that we are above a huge ball, this should be quite a sensation, an epiphany for many. An even more unique sensation will be watching that ball diminish in size as we ascend, dwindling to the apparent size of a baseball held at arms length by the time we get to GEO.

By the time our passengers arrive at GEO station they may be "viewed out" and ready for the numerous delights of zero gravity. We have seen the astronauts floating about the cabin, playing with their food and using handholds to move around, so we have a good idea of what that is all about. But what they don't show you is the hygienic apparatus and the shower bag. That might take some getting used to. Bedrooms won't have beds. Sleeping would be just a sheet pinned down all around so that you don't drift off, not exactly the cuddling and snuggling in a soft warm bed that vacationers often partake of. All the ball and stick sports would be fun and amusing. Without ballistic trajectories the unexpected results would take a lot of relearning. Three-dimensional versions would develop with zones marked out instead of nets or goals. Since bouncing off walls would be the chief means of locomotion, team sports would change to strange new forms. If we had the luxury of large open spaces, flying would be possible with just large stubby panels strapped to your arms.

Anyone spending much time in zero gravity will have to have an exercise routine to counter loss of muscle strength and tone and eventually bone mass. So after the newness of zero gravity wears thin it would seem that space proprietors would have to offer their clientele artificial gravity accommodations, for comfort if not for health. Some portion of the space hotel will have to be built to rotate. This will offer a whole new set of fun experiences, old or new games with projectiles will be very strange. The coriolis forces set up by the rotation will cause thrown objects to follow different curved paths when thrown in different directions from the same starting point. Receiving or hitting a thrown object will be just as mysterious.

Once there is ongoing construction taking place in orbit all of this strangeness will be duly reported back to the Earthlings via the Discovery channel and the horde of TV reporters that will want to visit. We could start a limited tourist trade early on. Since we are building large facilities in the middle of the Pacific Ocean, reportedly a nice place to visit, we could start a ground tourist business. This would include tours of the ribbon launch and support facilities, an IMAX theater for that Wow view of space, and a "space camp" type introduction facility, as well as the regular ocean resort and water sports attractions. This can accommodate thousands of people and a decent cash flow at more mundane luxury resort prices. From this group we can glean our space travelers, assuming that there is still a quantity of wealthy adventure seekers left after this self screening process. A mystique of "hardship chic" could develop around space travel and then all of upper society will just have to go.

Since we will be transporting workers we could build a climber version for paying passengers. To avoid restrictions we could sell club memberships and then we would not be engaged in "public transportation". Club members would get medical examinations, classes and movies. Start out at $1 million each and see what the traffic becomes. We might be surprised. As new ribbons are added, work the price down to $100,000 each until we are ready to support a full-fledged tourist service. This may be far short of a tourist industry but together with the ground activities it would generate a cash flow that will be worth the effort in support of the Elevators.

In the future, when we do get into regular tourist traffic we may be able to setup separate facilities for tourists. That 15 day round trip to GEO may well be a detraction to otherwise disposed clientele. One solution to this problem could come about when ribbon material drops below $16 per kilogram, then a 500 ton capacity ribbon would cost less than $2 billion. We could then devote a ribbon to the tourist trade by hanging a hotel module at the 0.1g (14,000km) altitude, where the view of Earth is much better than way out at 36,000km and there is still a tenth of a gravity for comfort. Without zero gravity, this would double the number of people that could enjoy the trip. At one tenth gravity a 500 ton ribbon could support a 5000 ton hotel, but then it would be at max tension and no climbers could use the ribbon below that altitude. So we have to plan on using some combination that will allow climber traffic and support the hotel. One combination would be a 2500 ton hotel, allowing up to a 250 ton passenger carrying climber to shuttle back and forth. Depending on a lot of construction details – will we make it out of inflatable materials, or what – this size hotel module could support anywhere from 500 to 1000 people.

Trip prices would still be in the $6,000 range, (plus extra for the board and room) because we have to pay for the ribbon, not the length of the trip. This might be offset some by having rental space at GEO for satellite facilities. You could still use the ribbon at GEO where there is no load on the ribbon but it would have to be the kind needing only infrequent climber traffic.

There is no doubt that space travel will be a fantastic experience, and like the old safari's, not always comfortable; a true out of this world experience for the first time. Now all we need are new expressions for great things.

Chapter 16:
Energy and the Moon

There are many other nice places to visit and things to accomplish in space. We will continue with our exploration of these subjects in this chapter, but with a slight twist, we will refer back to a different question.

What would be the effect of <u>not having</u> the Space Elevator, on the world economy?

In general we can say that with Man's propensity to explore, we will eventually get out there and develop space, one way or another. And we have made our case that the Space Elevator will make this oh so much easier, cheaper and faster. But in the grand scheme of things does that matter? YES! During this century we will face a variety of serious, economic, political and environmental challenges. How we handle these issues will make the difference in whether our great-great grandchildren curse or praise us.

Energy

One of the ways often used to chronicle and measure Man's progress is through his use of tools. Another, and closely related measure, is his use of energy, a key feature distinguishing humans from other life forms. Now tools, once the archeologists find them, are at least definitive, energy usage, being historically anecdotal, is less reliable but none the less pertinent, for it describes the standard of living and life style of an age or culture.

Ever since Man first learned the controlled use of fire he has been on a never-ending quest to acquire and use more energy, in other words, to improve his standard of living. Indeed, the rate of growth of Man's progress can be marked out by the relative cost of energy to the rest of the economy. One of the best ways to judge the degree of a civilization's advancement is to measure its energy usage per person; it is a measure of a culture's wealth. This is particularly noticeable in the last two centuries. Access to relatively cheap energy has led to rapid economic progress, which has led in turn to acquiring more and cheaper energy. Our standard of living today is directly tied to the availability of inexpensive energy. In fact, we measure and celebrate our success in life by how much energy we can use, as a leisurely drive through your nearest wealthy neighborhood will attest. Money, of course, is the direct way of keeping score, but, just think about it, almost everything we do with greater wealth is to consume more energy. Even the money some of us don't spend is saved, becoming part of the financial infrastructure that is used by others to create more wealth using energy.

Now comes a variety of periodic energy shortages and prognostications of greater shortages in the future. And what is always the first thing suggested as a solution? Conservation! Save energy, don't use your well-earned wealth, deny yourself, sacrifice your pleasures and needs to the greater good. (This is particularly true of the Elite slandering large cars and SUV's, while loving the Internet, which recently has accounted for the largest increase in energy consumption.) This is entirely wrong headed. Sure, we could use less energy (using it more efficiently would be good), but that is strictly a short-term solution. The feel good ecology that promotes that approach is misplaced, in the end it means our advancement, individually and as a culture, will grind to a halt and with it our wealth and then our standard of living. Energy bestows wealth. For those few that actually do want less wealth, that can be arranged.

Our jobs, as members of a wealth producing culture is not to flagellate ourselves by restricting our use, but to go out and find more and cheaper energy; if not ourselves, at least to promote that process. The key to our society is energy and the resources that energy can provide. Given enough energy we have no resource problems, we can go out and get, or make, any resource we need. As stated above, Man must continue his "never-ending quest to acquire and use more energy, in other words, to improve his standard of living."

The next Mother Lode

Having thus led the cheer for progress we must pause for a sober moment and make a point as plain as possible. We will run out of Hydrocarbons, and a lot sooner than most people think. Second truth. Lag time, that is, the time it takes to start whole new industries and their supporting infrastructure, is going to be a problem. For new industries to replace current industries could be 20 to 30 years, even with a deliberate push to do so. If we don't get a running start on several types of solutions, we could suffer serious damage, the collapse of fuel supplies, and thus the collapse of the economy, before new sources are online. Specifically to our area of interest, if we don't have the SE by the time we run out of oil and gas we will be in a world of hurt, indeed, a world that hurts, unable to get workers to the workplace much less build a ribbon.

Think of the problem this way, suppose that a report came out, (a true report, however that would be determined) that said that we would run out of oil and natural gas in ten years, what would you suggest we do about it? You are not allowed to stone the author, but must come up with alternate energy sources and get them working in ten years.

Conservation?   Forget it, that only stretches 10 years into 12, in fact we may have to spend a good deal of the oil that is left to build those alternate sources.   And besides, nobody seems to have thought that since you have to use resources at some rate, however diminished, you will eventually run out anyway, so why be miserable for all of that longer time?   Daunting isn't it?

We have plenty of energy, we are swimming in a sea of energy, it is all around us.   The problem is energy density, how concentrated the source is.   Right now we are using the "backup reserve" of the sun.   Plants and animals on Earth grow by converting sunlight into mass.   The Earth saves up this mass, cooks it down to a nice concentrated form and stores it as fossil fuels that we dig up and burn.   There is no doubt that we will eventually run out of these resources.   The only question is whether we will run out in 30 to 40 years by using them as fuel or in several thousand years by saving what is left for use as petrochemicals; the lubricants, plastics, fibers, solvents, pharmaceuticals, fertilizers and other chemical byproducts that are as vital to world commerce as fuel.

All the other sources of chemical energy (burning) are not dense enough sources for modern needs; we can't run cars by burning wood. So, rather than sit here on our big fat duffs and whine and complain and wring our hands, "Oh dear me, we are running out of natural resources, we must stop using them", the real answer is to go out there and get more resources.   What about "renewable energy" sources?   That is a misnomer, they are all secondary, or indirect, Sun energy, you might as well go to the source and cut out the middleman.   In addition, they are one or more of the following: not very dense, not local, not portable, not ubiquitous or are otherwise not engineeringly feasible for supplying any large portion of our energy needs.   To the extent that any of these can be made economically viable, even on just a local basis, go to it.   But do keep in mind that the return per dollar invested is likely to be very small and also is likely to detract from the effort needed to make much larger progress with the global methods discussed below.

Other than rearranging the outer electron shells of atoms (chemical energy) there are only two other methods of getting energy in a usefully concentrated form.   Direct energy conversion, the solar power satellites (SPS) we described in chapter 14, and mass conversion, our old nemesis, nuclear power, $E = MC^2$.   We noted that while the Space Elevator's lower lift costs make the SPS's possible, the cost of the materials still meant that we would not use this source until the cost of energy had already suffered a significant increase.   Thus SPS's become a "stop it from getting any worse" solution.   Could the other method, mass conversion, offer earlier and/or lower cost energy supplies?   Mass conversion comes in two flavors, fission, splitting the heaviest atoms into lighter ones and fusion, combining the lightest atoms into heavier ones. Fusion, the joining of atomic nuclei releases 10 times more energy than fission; Nature seems to clearly favor togetherness. (Fission converts 0.1% of mass to energy, Fusion converts about 1.0%, plug those numbers into $E=MC^2$ to get energy potential, from a "fuel".)

Fission we have, we know how to do it. From an engineering/science point of view it is cheap, safe and fast to add capacity. Currently, 22% of US electrical supply is nuclear. This is not a "renewable" resource, the world supply of uranium (U) is about 1.85 million tons. As used today in Light Water Reactors (LWR) that is a 200 year supply if use continues the same. If ramped up to takeover hydrocarbon use, it is only about a 46 year supply at today's energy levels, which would, in turn, be diminished to much less than that by the growth in energy demand. It would seem then that fission nuclear is only a temporary reprieve. We need something with a longer lasting supply and hopefully a greater energy density.

(By the way, LWR's are a very poor way to use uranium for power, it is not the engineer's way to do it, it is the political way of doing it. The breeder reactor method makes plutonium from uranium, internally, while power is being generated. The plutonium can then be refined out and used in another reactor to make more power. The same quantity of uranium is then worth 47 times as much energy, a 9,400 year supply at current use, a 2,300 year supply if used to replace hydrocarbons.)

We have that better supply and greater density in fusion reactors. The trouble is, we haven't perfected it yet, but it does have better press, everybody agrees it would be just so wonderful to have. If so, then why are we only spending $250 million a year on developing it? Disney spends more than that on a couple of movies.

There are two basic types of fusion. Magnetic Confinement Fusion (MCF), the Tokamak, torus shaped vessels that heat and compress a large plasma mass using very large superconducting magnets. The other is Inertial Confinement Fusion (ICF), where a fuel element (pellet) is imploded by various methods. As noted, progress has been slow, but recently several devices have reached sustained breakeven; that is, the thermonuclear energy output is equal or better than the energy input it took to get it started.

Magnetic confinement, the Tokamak, the first and longest under development has suffered some disillusion about its ability to take a major role in power generation. For one thing they need to be very large to get a decent amount of power out, and even then they are only about 35-40% efficient. To be more efficient it looks like they will need warmer (higher operating temperature) superconductors, which we have discovered but haven't perfected as yet. Also, it turns out that they do have a radiation problem. Not like the fuel rods of uranium or plutonium in a fission reactor but rather a slow buildup effect of running the fusion process. Its fuel is either, deuterium (D) ["heavy water"] or D and tritium (T) ["very heavy water"], both isotopes of hydrogen. [Hydrogen, (H) is one proton (p), and one electron (e), or (pe). D is one p, plus the same electron, plus one neutron (n), or (pen). T is still one p and one e but two n, or (penn).] D+D or D+T reactions are the easiest fuels to "burn" (fuse), but they both produce a free, high-energy neutron. The rest of the reaction products are electrically charged and thus are controllable by the magnetic fields of the reactor. But neutrons don't carry a charge so they fly off wherever they please. The reactor chamber has heavy side walls, built to stop them and extract their kinetic energy as heat to run the electrical turbines. Over time, the side walls (and other components) are turned into radioactive material by this constant bombardment. (It can also change the physical properties of the materials, so they no longer do their job properly.) This makes for a large maintenance problem, refurbishing the reactor from time to time and more radioactive, handling and disposal problems.

Deuterium (D) is easy enough to get, it's one part in every 6700 water molecules, so there are several trillion tons of it out there. But tritium (T) has to be made by bombarding lithium with high speed neutrons, and this energy cost of harvesting the T has to be added into the energy cost of lighting off the fusion reactor as against its energy output. (By the way, T is radioactive with a half-life of 12.3 years, so it doesn't store well. T is used to build hydrogen bombs, which in turn do not store well. If H bombs are not periodically "recharged" with fresh T they will no longer be H bombs. It has been reported that the U.S. is the only country currently doing this. It looks like the "cold war" really is over! The reason that we use the heavy hydrogens for reactors and bombs is that it is far easier to fuse them than using plain hydrogen the way the stars do it; about 100 times easier.)

The other method, inertial confinement fusion (ICF) has better prospects, and they are more efficient at 60-80% energy conversion. ICF which is the high-pressure implosion of a small localized mass, as opposed to the volume of the chamber mass of MCF, uses more precisely and effectively directed energy input. Actually there are two types of systems, the laser "sock-it-to-em" method that most of you have heard about, and the electrostatic "suck-you-into-my-parlor" method. The laser method is still quite a huge machine, perhaps using as much if not more hardware than the magnetic (Tokamak) method. While the "Inertial Electrostatic Fusion" (IEF) method looks like it has a good chance, once developed, to be quite small; recent experimental models are only 2 feet across.

IEF works by having two concentric spheres, an outer positively charged grid and an inner negatively charged grid. The warmed up, pre-ionized fuel (charged particles, so that they can be driven by electric fields) is sent into a surrounding chamber where it is accelerated by the two grids to a crushing rendezvous at the center. This arrangement (without going all the way to sustained fusion) is highly useful as a commercial source of high-energy neutrons or protons, for producing such things as medical and industrial isotopes. But for power production, the inner "grid" is replaced by an injected cloud of electrons, a physical grid would be melted.

In addition to being smaller and more compact, an intuitively more elegant design, the really big deal is that IEF's and cousins, can use new fuels that don't produce damaging neutrons. Thus no heavy shielding, you can put them anywhere. This next generation fuel is D + Helium-3 (^{3}He) an isotope of helium that is short one neutron. (Two protons, two electrons so it is still helium, but only one neutron.) This reaction produces one p and a regular He (4He if you will) and a nice strong energy pulse of 18.4 Mega-Electron-Volts (18.4MeV) a little bit better than the D+T reaction. The p is a charged particle so it can be controlled as needed and the 4He is not an obstreperous particle, it just bounces around until it gets lost. These fuels are highly efficient, the reactant particles fly out of the reaction chamber at high energies and can be converted directly into electricity by electrostatic plates at 70% to 80% conversion rates, without the bulky heat exchanger plumbing or steam turbines and with no residual radiation effects. Wow, what an ideal setup, stick power plants anywhere, in the city, on ships, "Happy days are here again. . ." well not quite, ^{3}He is very rare, barely enough to do experiments.

Here is where the Space Elevator comes in; the nearest supply is on the Moon.

Fortunately we have this Moon that has been so thoughtful to have been collecting ^{3}He from the solar wind for the last 4.5 billion years, saving up more than a million tons of it just for us. One kilo of ^{3}He is worth 157,480 barrels of oil. By the time we get around to mining the Moon it will be because oil is running low and it is high priced. At only $50 a barrel, (it could be a good deal higher by then) one ton of ^{3}He is worth 7.9 billion dollars!! We only need 32 tons of ^{3}He a year to replace the fossil fuels the US uses for electrical generation. For our own use that is a market value of 252 billion dollars a year, bring back three times that to sell to the rest of the world and it's worth over 750 billion dollars a year, more than enough to get a small mail order business going.

Even with the staggering future value of ^{3}He as an inducement, it doesn't have that value now, and without shortages and the new fusion reactors perfected it may not for some time. But with the SE we can afford to go there, explore and assay the Moon a lot sooner and a lot cheaper than when an oil shortage panic would push us there. (Remember we have to deal with lag time, or pay dearly for falling behind.)

Without the SE

While the prize is huge, what would it cost to attain it? Without the SE we would have to go there on rockets and that would cost an awful lot. A comparison of mass ratios tells the tale.

The Space Shuttle (SS) starts out at a gross weight of 2041 tons and delivers a 25 ton payload to LEO, a mass ratio of about 82 to one, very inefficient. It also costs $10,000/kg, if you could double its use rate per year. But the SS is actually delivering the Orbiter with payload to LEO or 109 tons, now the mass ratio is down to 18.75. And you can also add back in the external tank, 30 tons, which could go into orbit along with the Orbiter except that they want it to reenter to get rid of it. That makes the true lift to orbit capacity of the SS 139 tons, a mass ratio of 14.7 to one. This tells us that an advanced lift vehicle like the Single Stage to Orbit (SSTO) vehicle they are trying to design won't be able to do much better than a 13 or 14 to one mass ratio, to LEO. This also puts the cost at $1,700/kg, about the best we can hope for with rockets, even if we build dozens of them.

Now that we have a payload in LEO, we still have to get it to the Moon. The ΔV from LEO to the Moon (trans-Moon injection) is 3.2km/s, and an additional 2.8km/s for orbit capture and landing on the Moon for a total of 6.0km/s. The best mass ratio we can hope for with chemical rockets, assuming we develop advanced ships for this part of the journey, is 6. That makes the cost to deliver payload to the Moon, $10,200/kg.

The next question is how much equipment and manpower would we need to ship to the Moon to mine the ^{3}He in the quantities needed.

The University of Wisconsin has an extensive program on the subject of space resources, and have worked out a plan for mining the Moon in considerable detail. Professors Thompson and Kulcinski have come up with an engineering concept for a miner, they call the Mark-II. They see the R&D program costing 3.2 billion dollars, and the Mark-II miners costing $10 million a copy thereafter. The miners mass 18 tons, not counting the solar mirror heat concentrators for the heat extraction

process. The Mark-II is designed to work the top 3 meters of the regolith*
like a dredging operation, grinding forward at the snails pace of 23 meters
per hour. Still, that does eat up one square kilometer a year. Out of that,
depending on the richness of the ore, it should be able to produce a
minimum of 600 tons of volatiles a year.

H_2	201 tons
H_2O	109 tons
He	102 tons
CO	63 tons
CO_2	56 tons
CH_4	53 tons
N_2	16 tons
^{3}He	33 kg
Total	600 tons

That last, very small quantity, compared to all the others, is the
payoff, 33kg of ^{3}He is worth 7.9 million dollars per kilogram or 261
million dollars a year.

*[The Moon, unprotected by an atmosphere has been
continuously bombarded by meteorites, particles and high energy
radiation for billions of years. This has broken and stirred up the crust
until most of the surface is packed dirt, called regolith, to a depth of 3 to
12 meters. This has both enriched the regolith and made it easy to mine.]

The miner, built to go after the ^{3}He, fortunately produces good
quantities of most of the other valuable volatiles, enough to supply not
only the human needs but also the rocket fuel for the entire, Moon to
LEO, round trip traffic. This is an extremely important point, by
supplying these gases on site the mining operation contributes over $5
billion a year, (in avoided transportation cost dollars) to its own upkeep
that would other wise have to be supplied from Earth. If it were not for
this local supply the $261 million a year ^{3}He cash crop would not even
come close to paying its costs.

Now we have some idea of the size of operations. To supply the
world with ^{3}He energy, just for today's electrical power generation, we
need about 100 tons a year. This will take about 3000 miners at a cost of
$33 billion, shipping mass 54,000 tons. This will require the services of
between 5,000 and 10,000 people, depending on the level of automation
and robotics we can achieve. To support all this we will need two dozen
heavy lift (50 ton payload) SSTO's for the Earth to LEO traffic, from half
a dozen launch facilities in the southern US; cost, $10 billion to develop

and $1 billion each. A large space station, perhaps two, 4500 tons for LEO transfer terminal and another transfer terminal in lunar orbit, LMO, 2400 tons. A fleet of two dozen Moon transfer ships, plying between the Moon orbital station and the Earth LEO station. These would be nuclear powered, 1000 Isp, 20 ton engine platforms that would hook up to cargo pods sent up from the Moon by chemical rockets. A dozen or more of these chemical fuel heavy lifters, at 28 tons for the engine platforms, would lift the cargo pods, from the Moon surface to LMO. All in all, this daisy chain of transportation facilities would number some 60 to 70 vehicles, with a like number of cargo and passenger pods, costing in the neighborhood of $160 billion by the time you got it all into orbit.

This is as nothing compared to the total tonnage we will need to offload on the Moon. Using gross estimating techniques to approximate the various components, in addition to the 54,000 tons of miners, the support facilities will amount to some 26,000 tons, with the manpower and consumables supplies adding 82,000 tons. That's a total of 162,000 tons costing $652 billion dollars, FOB Moon. Earth based facilities will add another $100 billion and if you add on a 10% contingency for things we have missed, you have a nice round total of $1 trillion. These cost estimates assume a great deal. They assume good management. With poor management, which might be characterized by a low level of commercial and corporate involvement, these costs could easily be two or three times higher. They assume that we will be able to build a second generation SSTO with a decent lift capacity (50 tons). That by this we can get lift costs down in the $2,000/km range and that the cost of space hardware will respond to economics of scale so that we are not spending $160 million for one air lock door as recently happened on the ISS.

Fortunately this does not have to be spent all at once. Indeed, it couldn't be spent all at once. The first few years will be slow because we have to build up the mine's production of fuel that is sent to LEO to support the outbound traffic. It is like a feedback loop. The outbound traffic can't add men and mine equipment very quickly until there are men and mine equipment to supply the rocket fuel. The above totals represent the requirements to build up to full, world supply, production in 20 years, counting from the first manned base, which implies another 5 to 10 years of research and ship building before the start. And this is the fast schedule such as we used with the Apollo program, and thus a good deal of that money would be spent up front, the "start up costs". Even with the shortages that would be motivating investments of this size we could not move much faster, because of lag time on Earth. It will take that same 20 years and hundreds of billions more to build out the all-new infrastructure of nuclear fusion power reactors to use the fuel coming from the Moon. A heavy financial burden, but not for long, cash flow of $50 billion a year in revenue can be expected within 5 years, $100 billion in 7 years, $250 billion in 10 years and at full production, $758 billion revenues a year. It will be cash flow positive 5 years from the start of mining. Startup capital paid back in the 11th year, $6 trillion in total revenues in the 20 years, a fabulous return on investment.

Even with the prospects of such a happy outcome, how long will it take for Congress to warm up to the idea of funding a project at $100 billion a year for 5 or 10 years? Wouldn't they come around a lot faster if we could show them that the Space Elevator could do the job for less than half that cost?

With the SE

The driving economic factors are changed radically both for energy consumption and materials. The massive fuel requirements for getting off the Earth and its 7.6km/s requirement are eliminated. The 6.0km/s used to go from LEO to the Moon is reduced to the 1.9km/s needed to land in the Moon's gravity well. The 6.0km/s used to return from the Moon to LEO is cut down to the 2.8km/s needed to escape from the Moon, plus a few more tenths of one km/s to maneuver to the ribbon. (Once you have the velocity to escape from the Moon, if you are pointed in the right direction, you can almost just drift over to the ribbon.) We no longer need to have the complicated system of space ships and space stations to ferry Moon-made fuel to LEO in order to get ships back to the Moon less expensively than supplying fuel from Earth. The ships now go directly from the Moon to the ribbon and back again, it is easier to supply fuel from Earth for less than it would cost to bring it from the Moon. This also means that we don't have to wait for the Moon mining operations to slowly build up to the level of being able to supply this fuel. We can send men and equipment as fast as we wish, fuel making is no longer a limiting factor.

This eliminates the requirements for the SSTO, the heavy-duty space stations in LEO and about half of the Moon ferry requirements, a total of 7,500 tons, and $140 billion. And as important, that savings is right out of the most expensive startup costs. On the Moon, we still need thousands of those heavy miners, so there is no equipment savings there. Of the other Moon hardware, men and supplies we save about a third, 33,000 tons and $104 billion but these are all high volume items, cheaper by the dozen. All together we save $225 billion in equipment and over $335 billion in shipping costs using the SE.

Our operating costs are also reduced. We don't need as much fuel refining and fuel handling equipment on the Moon, and no fuel ferrying launches. This also reduces the manpower needs for these activities.

Fewer ships in transit mean a corresponding lower manpower and handling equipment requirement on the Earth end of operations both in space and on the ground. Adding these savings into the mix and the bottom line is that the same 20 year mining project only costs one third as much with the SE as with ground to orbit rocket power. One of the limiting factors could be the rate of daily traffic. Kind of like the tides, you need to launch from the ribbon at a particular time each day when the ribbon is swinging around toward the Moon. How short is this period? How many inbound and out bound shipments can be handled in that short period will decide if we need additional ribbons to handle the Moon traffic. But by then, they only cost a couple of billion.

Perspective

Mining the Moon as just outlined would be, by far, the largest construction project ever attempted; and yet it would not hold the record for long, for we are just getting started. This would only replace fossil fuel usage in electrical power production as now constituted as a segment of our economy, a savings of about 25% of our current fossil fuel usage. In order to save all the other fossil fuels used in transportation, we will have to shift energy production to electrical power plants and then figure out a way to get it to vehicles. Whatever fossil fuel sources are left will have to be saved for non-burning uses, petrochemicals as mentioned above. (One irony will be that "fossil fuels" will no longer be fuels.) This means that the electrical power generating facilities will have to grow many times faster than the growth of energy usage in order to take over all the energy supply.

The huge advantage of portable power in the form of fuel in a tank, can't come from gasoline any longer. Yet we must maintain this portable power, fuel-in-tank, structure, for our economy is based on mobility. It would not take much of a reduction in our mobility to drive the world economy into a serious depression, one where the nightly news counts bodies. We must therefore develop other methods to have portable power. The only power sources available for vehicles would be electrical or chemical (but not from fossil fuels). The only way to make electrical power portable, this century at least, is with batteries or fuel cells. Batteries might be developed to the stage where they are good enough for low power, short distance usage. More likely, for anything resembling the power and duration of gasoline powered cars and trucks, fuel cells will be the thing.

The food chain then goes from solar and nuclear electrical generation, down the power grid to charge batteries locally, or to processing plants to make hydrogen for fuel cells which is then shipped to service stations to be dispensed to vehicles. This works nicely for ground vehicles, which would save a huge percentage of our fossil fuel usage. For aircraft we still need fluid chemicals to burn with oxygen in the air. This might be done by making synthetic fuels such as methane (CH_4); we don't really know as yet. These synthetics can be made using electrical power and sources other than fossil fuels. (Hydrogen is no problem, it is the carbon that we need, and there are sources, other than fossil fuels – we just have to work on it. However, it is interesting to note that, at full production, our little Moon operation, as a by product of the extraction process, produces over 120,000 tons of pure carbon a year.) Why not just use these handy synthetic fuels in cars too? We might, but the way it looks from here, there will be a price difference that will make fuel cells better for ground vehicles, whereas synthetics are most needed for aviation. What ever we come up with, the bottom line is the same; all energy usage will start with electrical generation.

Reducing all the different measurements of energy, barrels, cubic feet, pounds, etc. to a common unit of measure, watts, then the world currently uses, from all sources, over 4,300 GigaWatts (GW) of energy a year, equivalent to 438 tons of ^{3}He. If the economy continues to grow at an average rate we will be using over 13,000 GW, or 1,330 tons of ^{3}He a year by the year 2100. (To try and stop this growth would have serious economic consequences no one would ever accept, unless you live in the bush and don't know any different.) By then, all of that energy will have to come from electrical power generation, 14 times more plant capacity than we have now, so you better get busy.

By the next century our little Moon operation, that seemed so vast an undertaking, will have miners crawling all over the place, producing not some piddling 100 tons of ^{3}He a year but over 1300 tons a year. Daily traffic will require at least two space elevators for out going traffic and two for incoming traffic, perhaps more, and we will be looking longingly, over the solar system, for richer pastures.

Chapter 17:
Space Development

With a half dozen or more space elevators up and running, solar power satellites being built, Moon mining going strong and a ribbon or two on Mars we can say with some satisfaction that phase two of space development is a success. We can look forward to making it through the 21st century with prosperity and plentiful energy supplies. And then we find out that we aren't done yet. This level of development will just about make it to the end of the 21st century – and then what? It's on to phase three. Or what will more likely be called, in retrospect, the real period of space development.

One realization will be, with an ever increasing need for energy, that the supplies of ^{3}He on the Moon will not last forever and we might find, with future technology, that it could be cheaper to get it elsewhere. Another will be the realization that just as it becomes dirt cheap to send all manner of goods and supplies up the ribbon to space, we won't want to. Earth will need all the resources it has and will want to import more, not send it off to support those "Spacers". And that is only fair, after all, there is a lot more out there than there is here.

Here is just some of the resources that we can look forward to.

WEALTH 1

With 6 billion people on Earth
Your share of some key, representative,
resources of the solar system is:

		Per Person
HYDROGEN	34,000	Billion Tons
IRON	834	Billion Tons
SILICATES	834	Billion Tons
(i.e. sand, glass, etc.)		
OXYGEN	34	Billion Tons
CARBON	34	Billion Tons
Kilowatts, power*	64.56	Trillion

If you are not getting your fair share of these resources, please write your congressman.

(*64.56 trillion kW-hours of *energy* in one hour, 565.5 thousand, trillion kW-hr of *energy* in one year, etc.)

Space is indeed a treasure trove. Other than low-mass high-value products like computer hardware that will still come from Earth for many centuries, the start of phase three will be when we deliberately set about the task of making space self-sufficient. The question then is where and how will we get these resources?

Needs
Earth orbital:
Near Earth activities need everything, of course, since there is nothing there unless we bring it there. It will be no great strain on Earth resources, and no great cost to supply water, vital gases and food, so that will probably go on for some time. But all the other materials, metals, other construction materials such as carbon fiber composites and plastics we would like to get elsewhere as soon as possible. As mentioned in chapter 14, what Earth orbit needs most is construction materials for lab and factory accommodations to make high tech components that can benefit from vacuum and zero gravity.

Moon:
The Moon needs food, metals, carbon and plastic components. Construction materials should be aplenty and the mining operations should be able to supply their own water and vital gases as by-products, but that is a Spartan existence at best and as the Moon broadens its activity base larger supplies of water and vital gases will be wanted.

Mars:
Actually Mars has just about everything we need – just not right away; it will take some time to bring Mars online as a source of many resources. For instance, Mars is rich in metals but we don't know if they are in concentrations, like Earth, that makes for low cost mining. So we don't yet know when Mars will be a net exporter of some metals. What we do know is that Mars has an abundance of carbon, oxygen and iron in the forms of CO_2 in the atmosphere and Fe_2O_3 in the crust. Both of these compounds can be catalyzed out to their constituents by a hydrogen process that produces water as a by-product. We can't get greedy and drink the water because we need to break it down to reuse the hydrogen. This means that what Mars will need most for its development, early on, is hydrogen, which really means finding water, for the hydrogen. (Hopefully we will find subterranean water in many locations.)

One of the products of these processes is methane, CH_4 a fine fuel for rockets, vehicles or industrial use. Mars has been busy the last several billion years oxidizing everything that can be oxidized, so in addition to Fe_2O_3 we also have lots of other oxides like SiO_2 and Al_2O_3 and in somewhat lesser supply all the other common metals and compounds. From this we can see that what Mars needs is manufactured goods, tools and materials to build extensive chemical processing facilities and from these in turn, manufacturing facilities.

In the end Mars will be a major exporter of many resources and manufactured products, perhaps the most important being food. Other than Earth, Mars will be the only other economical (low energy input) food growing location in the solar system because, eventually, crops can be grown in the open. [See Zubrin, "the Case for Mars" 1996, wherein he notes that as Terraforming progresses plants can be started with as little as 0.7 psi atmospheric pressure under greenhouse domes and proceed to the outdoors as soon as the atmosphere is thick enough to maintain growing temperatures. In Mars high CO_2 atmosphere plants can grow three times faster than on Earth.]

Given its lower gravity well, enhanced by having space elevators, plus it being closer to everything further out, Mars will be a major supplier of food and manufactured goods to the rest of the solar system and an importer of any raw materials it may lack. It will eventually become the trading cross roads of the solar system. In the meantime, to speed up the terraforming during the development phase of Mars, it could use extra greenhouse gases to help raise the temperature and the atmospheric pressure. And, until we get the north polar ice cap (water ice) flowing, a little extra water wouldn't hurt either.

Outer system:

For everything out past Mars we will have to bring everything with us, for this century at least. Later, we will build processing facilities in the outer system to derive all our vitals needed locally and to make intermediate products, less bulk for shipping to the inner system, and later still, manufacturing facilities for goods used in the outer system.

We can summarize by saying that all space activity will first of all need consumables: water, the vital gases of human habitat and rocket fuel. Once this need is satisfied we can go after the metals and minerals out there.

Resources:

Other than Earth, the best supplies of these resources are the asteroids. The asteroids really do have everything we need. The professional astronomers have an impressive array of classifications for asteroids (to include comets), but for our purposes we can simplify that list considerably. (For an excellent presentation, and more than you ever wanted to know about asteroids, see Dr. John S. Lewis's books: "Space Resources", "Rain of Iron and Ice" and "Mining the Sky".) There are 4 basic types of asteroids that concern us. Ice, stone, iron and a complex type called carbonaceous. Asteroids can have any mix or combination of these materials, it's what they have most of that makes them one type or another.

The icy asteroids contain our first love, water ice, as well as several other frozen volatiles, mostly ammonia (NH_3), and methane (CH_4) which are great for fuels and also are great greenhouse gases for Mars. (As you may have guessed, for the next century or two we are not going to concern ourselves with what Mars smells like.) The stone asteroids are made up of a wide variety of silicates and oxides of a dozen or more of the most useful elements in the periodic table, just the thing for supplying an industrial base. These are the stuff of which rocks are made, or would be if you had a planet to cook them in. This is primordial rock, other than gas, the oldest stuff in the solar system, and in granulated form, easy to get at and process. Other stony asteroids are made up of real rock, pieces of the current planets blown into space by massive impacts of their prior cousins (once removed!).

The irons, of course, contain mostly iron, but a good deal of the other metals as well. Many are actually natural stainless steel with mixtures such as 92% iron, 6% nickel and 1% cobalt. Just cut it up in the shapes of knives, flatware and utensils and we are in the import business. The last one percent is typically made up of the platinum group of precious metals, which are as valuable as all the rest of the asteroid. Lewis makes a loose assay of a typical iron asteroid, 2km diameter. It would mass 30 billion tons, the iron and nickel would be worth $8 trillion, the cobalt $6 trillion and the platinum group another $6 trillion for a total of $20 trillion. As if these astronomical (pun intended) sums were not enough, it might actually be worth more than that. It is not clear what values Lewis is using for the individual items in his evaluation, but the one percent for the platinum group would come to only $20 per kilogram. A more realistic average value would be more like $200/kg (and some of it would be worth over $2,000/kg) for a value of $60 trillion just for the platinum group and you can forget about the rest of the asteroid.

This is the type of science fact that persistently feeds the familiar Science Fiction theme of the "Belters", the transposition of the American Wild West to the asteroid belt. Where wild and woolly individualists prowl the asteroids for that one "find" that will make them eternally wealthy while they battle the big meanies for the right to be free. While that may be wildly unrealistic, given what robotic explorers could do, it is true that the last one percent would be well worth the transportation costs.

But, as is often said about earthly treasure, you can't eat it. So, for many aspects of space development, the carbonaceous, or C type asteroids may be of more importance. They come in a wide variety of mixtures but basically they are a collection of water-soluble salts, carbon compounds, ("organic" material but not from "organisms") and clay materials. In other words, all kinds of goodies to a chemist, from which we can get most of the compounds we need. C type material can be the whole asteroid or as large parts of stony asteroids, and it is plentiful, making up from one third to as much as one half of all the asteroid and comet material out there.

The trouble is, these different kinds of asteroids are scattered all over the place. It is estimated that there are 40,000 asteroids, just in the belt, one kilometer wide or larger and over one billion down to 40 meters wide. And yet the area they cover is so vast that if you were on an average asteroid of the belt you most likely would not be able to see a neighboring one, even with the assistance of field binoculars. This means we will have to journey far and wide and use different collection and processing techniques to get all the goods we want.

The main asteroid belt lies between 2.0 and 3.3 AU, (Mars is 1.5 AU, Jupiter 5.2 AU). A one-way minimum energy trip from Earth will cost you one year and 8.3km/s ΔV to the inner band and two years and 9.7km/s ΔV to the outer asteroids. And almost half of that energy must come from engines and fuel, see note below. So, you can see that with several years to "work" the asteroid these are going to be long term projects. Even with the ribbon available to give you, up to 2/3 of the outbound ride* almost free, we will need large ships for all the equipment, man-power, rocket engines and fuel supply to maneuver in the belt. Bringing things back from the belt will require large engines and huge fuel supplies, hopefully the fuel can be supplied by the asteroids. We will need time to build up our space going facilities and strong incentives, to go a-mining in the belt. There are, however, several ways to minimize these problems.

[*The SE or any launch method can only impart the outgoing ΔV, (4.3 to 5.0km/s in the case of the asteroids) for the distance to be traveled, not the orbit matching and maneuver ΔV at the end of the trip. This is large for asteroids and minimal for trips to planets where a variety of methods can be used to change speed and vector for a "capture" that saves fuel. That is why it would be so nice to find ways to apply fuel saving elevator technology to asteroids, as we will see shortly.]

Not all of the asteroids are in the "belt", not even most of them. There are, for instance the two Trojan clusters that lead and follow in Jupiter's orbit by 60 degrees (the L4 and L5 points of Jupiter). But more importantly there are many asteroids that do their work in orbits that come close to Earth, called Near Earth Asteroids or NEA's. Although asteroids impact the Earth all the time, some 10,000 tons of them a year, it is only the large ones that concern us.

A 10m asteroid has the potential explosive force of thousands of tons of TNT and they frequent Earth about once every year or two. "Potential" being the operative word, usually no one notices because they don't reach the ground, breaking up and dissipating their power high in the atmosphere. But larger ones would be noticed, especially if they reach the ground and especially where people live. The Tunguska, Siberia blast of 1908 was caused by a meteorite only 43m in diameter that blew up from the massive air pressure of its entry just short of reaching the ground, a perfect "air burst" that would have leveled any city had it been over one. Meteor crater of Arizona was caused by a meteorite only 150m in diameter that did reach the ground, devastating most of the area that would become Arizona 50,000 years later, so there was no one to complain. A one kilometer asteroid, reaching the ground, would take out a continent and make a "nuclear winter" for the rest of the world, and they come around about every 300,000 years.

The Spacewatch program was started in an effort to at least understand this threat and can detect NEA's down to 10 meters. To date, some 400 have been identified and from this sampling it is estimated that there are 2,000 NEA's one kilometer or larger, with tens of millions down to the 10 meter size. It is only in the last several decades that NEA's have gained considerable importance by the realization that this is the main supply of asteroids that impact with the Earth and have caused great damage, such as wiping out the dinosaurs 65 million years ago. It is now known that large impacts, powerful enough to wipe out most of the life on Earth, have happened repeatedly in Earth's history and are likely to happen again. So far, these impacts have been a good thing, tilting the course of evolution in our favor, but enough is enough, no more tilting is needed.

Right now there is literally nothing we can do about the big ones that may be headed our way, other than finding out more about the traffic out there. But as we gain space capabilities, a strong argument can be made that one of the first responsibilities of a civilization that gains access to space is to protect itself from extinction. And not just to save the Earth. Once we are active in space, with installations in orbit, on the Moon and Mars we present multiple targets, a "target rich" environment from the asteroids point of view. It would be very worth while to know what is coming, when and where. This is the motivation to put in the time and energy to find, then go to the near Earth asteroids.

With the space elevator we will be deploying large astronomical observatories to orbit, some of this capacity can be devoted to the Spacewatch program, mapping asteroids down to one meter in size. From orbit all wavelengths can be used, light, infrared, radar would work great from space. After we identify the ones that pose a threat we can send out robot probes to explore them in detail and mount a beacon on them for easier tracking. Better tracking will allow us to predict its orbit many years in advance, a long lead-time being one of the things we have to have should we need to take definitive action. If we want to change an orbit we may have to start doing so 10 or more years in advance of when it would impact. We also need to know the geology of the asteroid, its size and shape as well as its physical make up, in order to know how to move it or to disperse it. And when defensive action is needed a manned expedition to the asteroid will most likely be needed.

Once we have done all this, to keep Lloyds of London happy, what have we got? We have got all the preliminary work for mining expeditions done, requiring little more energy than a change of mind to go a-mining. The other problem, of time and distance, is at least partly solved. NEA's are a lot handier than the asteroid belt. Lewis has calculated that outbound ΔV's of 3.3 to 5.0km/s will get to most of them (less than going to the moon), with inbound ΔV averaging 1.0km/s if you chose your return point near Earth. This is less than half the round trip ΔV mentioned above for the asteroid belt. These would still be long trips, probably 3 to 5 year tours of duty for the people working the mines and perhaps decades for the large bulk transporting of products to various destinations around the solar system. These are clearly large scale projects, both time and capital intensive. The last problem is, where would the money come from for such a long-term pay off? Capitalism to the rescue. Since we would have a known asset, a well explored asteroid with probably more physical data than a good mine on Earth, we can sell shares and/or development bonds in this property. "p'ss, hey Mr., wanna buy an asteroid?"

In developing asteroids our carbon nanotube ribbon material will be a major working tool. In this case as rope, lots of rope. With no gravity worth mentioning, working an asteroid is going to involve tying things down, lashing the ship and equipment to the asteroid. Anything that uses a pushing (or pulling) action, drilling, digging, will need strong tie downs. The sheet materials for inflatable habitats, domes, solar power panels, solar sails for moving asteroids or payloads and the ribbons to them, will all benefit greatly from a super strong, lightweight material.

In some cases we may even be able to build an elevator on an asteroid. As far as we know everything in space rotates. If the rock is not spinning too fast, and the plane of rotation is in the right direction, we can run a ribbon out as far as we need and use it to send ships and payload back to earth. Knowing the particulars of our asteroid before we go we can build and send out an elevator on a spool. Counterweight material can come from the asteroid and the power beaming lasers can be mounted anywhere, depending on how we orient the solar panels. We only need to power the climbers to come down the ribbon, a climber on the bottom of the ribbon just needs a little push to get it started up the ribbon. With little gravity, all the force on the climber is centripetal, outward force. What it needs is brakes.

As an easy to visualize example take Earth's GEO conditions. At 42,163km from the center of the Earth our orbital velocity is 3.07km/s (rotation period, 1 day) and we are not going any place because of Earth's gravity. But with the small mass of an asteroid in place of the Earth, all of that 3.07km/s is useful velocity to go some place. With a standard asteroid rotation period of six hours a ribbon one quarter the length of Earth's will give us the same useful velocity. So, we might expect to get decent velocities out of asteroid ribbon lengths of 5,000 to 20,000km. And they would be very light weight, or rather low mass, in the 10 to 40 ton range, not the hundreds of tons of Earth ribbons. As an example, an asteroid with a 0.25 day spin rate would have the same 3.07km/s velocity at the end of a 10,541km ribbon as we do at an Earth GEO altitude, and without having to overcome gravity, it would all be useful trip imparting velocity.

Ribbons as short as 7,000km would impart 2.0km/s ΔV, and 12,000 would impart 3.5km/s ΔV, good enough to cover the range of return trajectories from most NEA's to Earth or Mars destinations. To handle 100 ton shipments such ribbons would only mass about 10 tons. While this is small enough that we could carry a generous supply of ribbon along with us as cargo on our mining ships, it is actually too small. With that mass it would not have enough angular momentum to prevent the ribbon from slowing down when being climbed and wrapping itself around the asteroid.

With just a plain ribbon it will take 11 days for a 100 ton load to climb to the end of a 12,000km ribbon without adverse deflection from the vertical; way too slow. We don't need much ribbon length to get the velocities we need so what we need are large counterweights (CW), out at the end to get a decent angular momentum. With a 200 ton CW on the end to increase the angular momentum the climbing time is down to 1.7 days and with a 500 ton CW only 19 hours.

We might as well get some use out of this mass and a habitat for the miner's would make a good CW. We need a good sized habitat – space for all the crew, lots of supplies, storage for equipment, pressurized bays for working on the mining equipment, etc. And it would have the very handy extra feature of artificial gravity due to centripetal force. The 7,000km ribbon mentioned above would have 0.06g or 6% Earth gravity, the 12,000km ribbon about 10%. Even a little "gravity" would make life much more comfortable for all concerned. Of course, now we need a stronger ribbon but you are still only dealing with ribbon masses in the 20 to 40 ton range depending on the size of the loads we intend to handle and the destinations. (24 tons for 12,000km of ribbon with load of 200 ton CW and a 100 ton climber.)

Even for asteroids that are not spinning in the right direction to work as launchers, it may still be worthwhile to take advantage of that spin by mounting habitats on the end of ribbons for the comfort and convenience factors. For asteroids that are not spinning enough to matter, we still might profitably use ribbons. We have to kill the overtaking speed of our arrival to match orbits with the asteroid. This would ordinarily require rocket fuel, but we might be able to rig some short ribbons that the incoming ship could snag on to. Just grabbing fast to a ribbon would wrap you around and crash the ship into the asteroid, so the technique would have to be to grab the ribbon about in the middle and let the ship slide, with resistance, outward on the ribbon as you were being swung around. Obviously this needs some experimentation.

Love bites from space

Somewhere, amongst all the asteroids are all the elements, so eventually we will find anything that we may be short of on Earth. The low(er) volume materials like precious metals, uranium, "rare-earths", "noble" gases and such, we can ship back to wherever in the solar system it is needed by ship. A billion dollars worth of precious metals is not that big a chunk. But for the bulky stuff, especially the frozen slush balls of water, ammonia and methane we will have to use other means. These materials are rather loose and may not take kindly to being pushed around; they would break up if we just mounted a rocket engine on them. One technique is to refine out the separate components, put them into very large tanks, let them refreeze (into a much more cohesive medium) and then push those chunks around with engines. The tanks can be made of carbon nanotube material, very light weight, very strong. Once the material is refrozen the tanks don't have to be all that strong for the journey. Heaters in the tank can draw off liquid for the engine's fuel. Another technique is to throw a net around the whole asteroid, to hold it all together, attach that to a solar sail via long cables and steer it to its new destination. All of the materials for this would, of course, be made with our ubiquitous carbon nanotubes.

These large loads would be the proverbial slow boat to china, taking decades in some cases to get to their destinations. At least they are not consuming any energy that we have to supply, even the onboard computers and robotic steering motors can be run on solar power. The answer for this time problem is to fill the pipeline. Get a long line of such towed cargo's moving, several a year; once the first ones start to arrive then you have a steady working supply. Do order in advance.

One of our most steady customers for this service will be Mars. As mentioned, the extra greenhouse gases that Mars needs to speed up terraforming can be supplied from the asteroids with small energy expenditures using solar sails. Find small slush ball asteroids of 40 to 100m diameter (one million tons) and get them headed toward Mars. Steer them to come in at a shallow angle to graze the atmosphere, slowing them down so that they go into long elliptical orbits around Mars. Each low pass through the atmosphere burns off large quantities of the ices and the friction slows the asteroid still further. The gases and water vapor dissipate into the atmosphere and any stone or gravel material falls out. Mars would have certain hard hat days, sorry about that. After several passes the asteroid is all gone. We may want to send someone out to retrieve the valuable nanotube sails before the asteroid enters orbit.

The wayward sister

What to do about Venus? Not much. At least not this century. Our sister planet is too darn hot and seems bent on staying that way. This is due to the now famous greenhouse effect, on a global basis, discovered in 1960 with our first probe. Carl Sagan quickly came up with a solution. Seed the thick clouds of CO_2 that envelop Venus with specially designed algae that would consume CO_2, combine it with water vapor, producing oxygen and multiply like crazy. Soon, from the thousands of tons of algae imported for the task, you have millions of tons working for you. As the percent of oxygen increases and the CO_2 decreases, at a certain level the atmosphere would become more and more transparent and huge quantities of heat would radiate off into space, breaking the greenhouse effect. Carl was a little too quick, it was later discovered that there isn't enough water vapor to do the job. It is short by about 100 million, billion tons! We could borrow Pluto's moon Charon (1200km dia.) and drop that on Venus and it would be about right. It would take thousands of 20km wide ice asteroids the rest of this millennium and most of the next to do the job.

But this objection to the Sagan-algae plan is based on the amount of water it would take to convert all of Venus's CO_2 via feeding it to the algae. This should not be necessary; any significant portion may do the trick. By the next century, if we were in the asteroid moving business, and if we could build carbon nanotube solar sails large enough, then two or three 20km wide asteroid might do it. This is, after all, a world-killer size asteroid, if there were anything to kill, 8 times the size of the one that did in the dinosaurs. The impact would be horrendous and it should blow a good portion of Venus's atmosphere out into space, produce a giant column of steam and churn the atmosphere. All of these actions cause the atmosphere to lose heat. In the Earth strike of 65 million years ago the immediate blast effect killed off a great many animals. But the secondary effects of putting up a plume of dust and debris that shrouded the world, cut off the sun and plunged the world into a multi-year long winter did most of the killing. The question for the astrophysicists is would the even larger dust cloud on Venus cut off the incoming heat and promote cooling or would it add to the opacity and insulation that is keeping Venus hot?

Now you drop in the Sagan-algae and with 4 to 8 trillion tons more water to work with it might just tip the balance to a more transparent, heat-losing atmosphere. At least it's worth thinking about. A variation on this theme is to use the Mars method. Have our large asteroid skim through the atmosphere and go into a wide ellipse such that it makes repeated trips through the upper atmosphere. In addition to stirring the atmosphere, while dissipating its water, it should blow or drag a good quantity of atmosphere out into space, getting rid of it. This lowers the atmospheric pressure, upsetting the heat balance and reduces the amount that needs algae conversion.

The only other way to cool Venus is to block the sunlight; it needs a parasol. Robert Zubrin has come up with a plan to build a Sun shade twice as wide as Venus and set it out at the Sun-Venus L1 position where it would stay, perhaps with a little nudge once in a while. His construct would be made out of 124 million tons of aluminum, 0.1 micron thick. At least this is less of an engineering project than moving 4 trillion tons of asteroid halfway across the solar system, but we don't think Bob will mind if we improve on his idea a little bit. The density of Al is 2.7g/cc, carbon nanotubes is 1.3 so right there we save 52% if built to the same dimensions. If 0.1 micron is strong enough to support the aluminum sheet then 0.01 microns ought to be easily strong enough for our nanotube sheet. That saves another 90% for a total savings so far of 94.8%. This brings the mass of the sunshade down to 6.448 million tons, a nice improvement.

There is one other savings that might work. We might not have to block all the light, it might be sufficient to block certain wavelengths such as the infrared, the major heat component of Sunlight. This would mean that the sunshade would not have to be solid, it could have tiny spaces between the fibers that would reduce the total mass of our sunshade. We can pass this question off to the optics people to work out the details but basically if the spaces were 0.7 microns* wide they should block, or at least diffract the infrared radiation. With nanotubes 0.0012 microns wide this, theoretically, would reduce the mass of our sheet by another 98.6%, depending on how the running, and cross fibers were constructed. Now we are down to 92 thousand tons, a far more manageable project. In fairness we should note that it may be very difficult to make such a fabric, but as a project for next century, what the heck, it may be easy by then and we have reduced the mass we have to carry around by a factor of 1348 to one by using our ribbon material. That is such a huge savings that almost any fancy fabric would probably be worth it.

(*It might be more advantageous to use smaller slits, 0.6 or 0.5 microns and block more of the light band but the overall density of our screen would only be increased a modest amount.)

One problem with this idea is that we have just built a very nice solar sail and it may be difficult to get it to stay in position. The idea would be to place it on the Sun side of the L1 point such that the gravity pulling it toward the sun just offsets its sail force. If this turns out to be unworkable then we could try plan B.

The other way to shade Venus is to put a wide ribbon, in orbit, all the way around Venus. The altitude of the orbit would be whatever is needed to clear the tops of the atmosphere and the speed of rotation would be whatever the orbital speed should be for that altitude, it would just be a very long, flat satellite that went all the way around. It doesn't matter which way Venus is rotating to do this and it has nothing to do with the geo-sync orbits that we are usually discussing. It would have to be very wide, we would have to do some research to find out what percentage of shade would be needed to have a decent effect, but let's say, at least a thousand kilometers wide. This would use far less material than the "parasol" above but the engineering problems of supporting it might be harder and add considerable mass. The main problem would be that the outer edges of the ribbon would not be "in orbit", they are not rotating over the planet's center of mass, only the middle of the ribbon is. This would make the edges want to move, to find an orbit. They would tend to bow up and then roll up towards the center of the ribbon, so some structure to supply stiffness will have to be provided.

If the very wide band (1000km) should prove to be a problem there is a variation on this theme we can try. Make several bands, of moderate width that we can manage, and set them to rotating at various angles to the mid-line facing the Sun. The first, and lowest one would rotate on the mid-line, (the "equator" to the clouds). The next one would be a little higher altitude and follow a path at say 10 degrees to the mid-line. The third one would be higher still and offset 10 degrees to the other side of the mid-line. Each additional ribbon would be higher and set off another 10 degrees either side of the previous one. Looking down on the sun side it would look like a swath of bandages that splays out to cover a good deal of the face of Venus. From the side they would all narrow down to the width of one unit as they pass over the same point and then diverge again as they go around the backside. A set of rings rotating inside each other.

Such planetary bands would not seem to need much in the way of strength. Theoretically there is zero tension along the path of its orbital movement, since all particles are in orbit and would remain right where they are if separated. But in practice we will need the nanotubes qualities of low density and strength in maintaining the width of the bands and for the stability of such structures we would induce tension by having the band rotate slightly faster than necessary for its altitude. This would also help solve the edge curling problem by having the center of the band tend to bow outward. We don't yet know the elasticity factors for large nanotube structures so this could be interesting engineering.

Once we have solved those problems and blocked some or all of the incoming heat the thermodynamic equilibrium that Venus is now in (surface temp 464C) will shift. The heat of Venus will now start to radiate away into space. The tops of the CO_2 cloud layers under the bands will freeze to dry ice and it will start to rain dry ice on Venus. The ice will not make it all the way to the surface, it will melt (sublime) part way down. But it will have cooled the layers that it passed through and each group of dry ice crystals will fall a little further until some reach the surface. This will take several centuries, and somewhere during the process the atmosphere will get more transparent to heat radiation and the dissipation of the ground heat should speed up considerably. Venus could still use a lot of water if we are going to make much use of it after it has been cooled so it would be a good idea to bombard it with a few moderate sized iceteroids while we are waiting for it to cool.

Round and round we go

By now we would have built inflatable and rotating, artificial gravity, structures in the hundreds of meters size. In the next century, having tackled "ringed-world" structures for Venus, we should have no trouble building "ringworld" structures. Not the Sun centered, planetary distance orbital objects of Larry Niven but the more modest form of 10 to 20 kilometer radius. You would spin it for "gravity", and dome over the inside surface like a bicycle tire on the inside of its rim, to hold air pressure. It is a cylinder slice of an O'Neill structure without the center portion being enclosed. As an example, take one at 10km radius (R) with the rim 500m wide; that's 62.8km around, and 31.4 square km of living space. Together with the arch of the dome, tens of meters overhead, there should be a feeling of considerable spaciousness. To get a comfortable amount of "gravity", say 20% (2 m/s^2), it would have to rotate at 0.01414 rad/s or one revolution every 7.4 minutes.

On the inner surface of the rim we could build anything we like, buildings, roadways, even parks and lakes as long as there was enough sunlight (filtered) to keep plants alive, say out as far as 2.5 AU. Let's propose a 2000kg/m^2 mass load on the inner surface, a "floor loading" as we would here for commercial buildings. Then at the rim's surface, gravity of 2 m/s^2 would generate a tension, T, in the rim, of 40 million newtons per meter of rim width (T=R*g*Mass/m^2). This may sound like a lot, but the carbon nanotube rim shell (manufactured by Space Elevators Inc. of course) would only need to be 0.62 mm thick to handle the load (including a safety factor of 2). The total mass of the shell would be a little over 25,000 tons but that is small compared to the 62.8 million tons of materials we could accommodate inside the rim. It would take a while to ship all this material into position to build the ringworld, even years, but it doesn't have to be done all at once. You can start with any useful width, maybe only 5 meters and keep building on to the edges. As tenants move into the new digs it would be best to send every other one to opposite sides of the ring, least the ring start to wobble – very distracting.

Now this would be an open ring or empty center design to show the basic mass ratios. How we get on and off it is another question. More likely we would build it with spokes, ribbons across the diameter to keep everything together while we are building. This would also give us access to the zero g center of the structure and allow for extensions of the spokes for launch ribbons. Notice the unusually large ratio of structure to useful load, one to 2500. This is an extremely efficient design, we don't have container to content ratios like that here on Earth unless you count wrapping 2 pounds of hamburger in Saran wrap.

What would we use such ringworlds for? Any place you need living, working, space and don't have a planet handy. As a commercial and manufacturing center in Earth orbit or the Moon's L4 and L5 points, where you process incoming asteroidal materials. As an operations and city center in the asteroid belt, serving a cluster of working asteroids. Or put one around an asteroid and do the processing and the manufacturing right there in a nice gravity environment plus living off of the volatiles. Extend two spokes out of opposite sides of the ring for launch ribbons and set the ring's spin in the plane of the ecliptic for efficient transportation to the rest of the system. The plane of the spin can be independent of how the asteroid is turning, thus making all large mineable asteroids ribbon-launch customers with low cost transportation. The only question is, how big will these things get in the next two or three centuries?

The outer system

Sometime in the next century we will get bored with just messing around with the inner system and we will strike out for the outer planets. There is so much mass out there, we ought to find something useful to do with it. Having worked the near Earth asteroids for a while, we will have shifted the emphasis of our asteroid activity to the asteroid belt beyond Mars. By now we will be Mars centric in our transportation thinking. Mars will be the center of action, with traffic outward to the asteroid belt for raw materials, inward to the Moon and Earth delivering refined materials and bringing back high tech goods. Having established the Lifestyles of the Rich and Famous on Earth, now spreading to Mars, our energy hungry race will have realized that the ^{3}He resources of the Moon, (only one million tons, 4 to 5 centuries worth) are a meager gruel indeed and that richer fields are needed.

The Moon is a great resource for ^{3}He because it is close and accessible with current technology, even if its ultra-thin concentration makes it laborious. Compared to what is available out there in the gas giants, it is meager indeed. Jupiter has over 500,000 times the Moon's supply, Saturn over 300,000 times, with Uranus and Neptune of proportional magnitude. The reason is simple enough, all gas planets have large quantities of regular helium and ^{3}He is along for the ride as a small percentage of 4He. So, the total quantity of ^{3}He in the gas giants is in the range of 1.4 trillion tons, which ought to last us a while; longer than the solar system will be livable. And you don't have to mine it, you just go out and scoop it up. Well, there is a little more to it than that, you do have to do some refining. The basic idea is to scoop up a tank full of the planet's upper atmosphere, a mixture of mostly hydrogen and about a third helium, depending on where you are scooping and separate out the ^{3}He. The separating of gases is cheap, the scooping is the hard part.

Harvesting ^{3}He from any of the gas giants is going to be a very large technical challenge, but with another century, or even two, to develop that technology it probably won't seem so hard by the time we get there. The basic problems are 1) distance, 2) deep gravity wells, 3) radiation and 4) we can't mount a SE ribbon on the planet – no land to anchor it. We will speculate on how we might handle these conditions and what help we can get from the SE. This will just be an outline of the general concepts, the detailed modeling we can leave to another time.

The Gas Giant Planets: (1 Astronomical Unit (AU) = 150 millionkm)

	Ave. distance (AU)	Years Travel	Planet Diameter (km)	Escape Vel. (km/s)
Earth	*1.0*		*12,756*	*11.2*
Jupiter	5.2	2.73	142,900	60.5
Saturn	9.54	6.05	120,900	35.2
Uranus	19.2	16.05	50,100	22.0
Neptune	30.0	30.5	48,600	24.0

[* Travel times are based on minimum energy Hohmann orbits for cargo to/from Earth]

Those travel times are awfully long, for people travel we will have to do something about it. Even with the SE, powerful engines will be needed both for the extra speed, and because of that extra speed, to slow down at the other end. Other than exploration, which we can do in detail this century thanks to the SE, manned development of the outer system won't happen without additional technological developments and until we are firmly established in the inner solar system. We can assume that in the coming centuries we will at least have nuclear fission engines, of 1000 Isp or better, in widespread use. These will be able to refuel on any of the readily available volatiles – the natural gases including water. We don't have to make chemical burning fuels for nuke engines. Hopefully we will also have fusion engines of 10,000 Isp or better using the ^{3}He that we are going out there to get. With fission engines we can cut the above trip times at least in half and with fusion to as little as one quarter to one fifth the time. Still, it will probably be better to break the trip up into smaller chunks by stopping off at Mars on the way. While Mars only takes a small chunk out of trips taking years, it would be an advantage because in the Mars gravity it would be easier to put up a really heavy, extra long ribbon to sling the rather large ships, with large ΔV's, that such long trips will need.

Jupiter

As can be seen, Jupiter is the closest ^{3}He supply at 5.2 AU from the Sun but the most problematic in the other areas. Jupiter's deep gravity well has an escape velocity of 60km/s from the surface of the clouds, compared to 11.2km/s for Earth, where without the SE we presently need two stage rockets to get to escape velocity. 60km/s is way too much for any chemical rocket or even our first nuclear engines at twice that efficiency to achieve if we had to start with no speed, as in floating in the Jovian atmosphere. So the trick will have to be, don't stop. One plan is to send a ship on a long elliptical orbit that skims into the upper atmosphere, scoops up tons of gas and then escapes out the other side to rendezvous with the refinery.

This is a lot more complicated than it looks. Because of the strong radiation fields ships operating near Jupiter should be robotic with heavy shielding for its key components. The ship will lose speed as it scoops up gas, it is accelerating the payload from zero to its speed, and the ship's momentum gets divided over the larger mass. The engines will have to be strong enough to make up for this loss. It would be interesting if the mass of fuel used was about equal to the payload we are taking on board. If we are clever with our plumbing system we might even be able to use some of the same tanks, saving a lot of redundant mass. All in all, a complex mass/energy management problem. Let's see if we can set up conditions that would make this workable.

The Galilean moons

	Orbital distance from center of J.	Orbital velocity	Orbital Period*	Moon diameter
	Km	km/s	Earth days	Km
Io	420,000	17.4	1.77	3630
Europa	671,000	13.8	3.55	3140
Ganymede	1,070,000	10.9	7.16	5260
Callisto	1,884,000	8.0	16.7	4800

*Note that for all moons the rotational period takes the same time as the orbital period.

The best place for a base of operations would be on one of the moons. Of the four large Galilean moons Io, the closest, is too volcanically active and too close to Jupiter's radiation. Next, Europa - the moon we would most like to have because of its ice-over-water ocean - is still in the outer edges of Jupiter's strong radiation belt similar to our Van Allen belt. So that leaves us with Ganymede, the largest, and then Callisto only slightly smaller. (Of the four, only Europa is Not larger than our Moon.) Both are outside the major radiation zone, have solid, inactive surfaces, and are thought to have good quantities of water ice and other ices, excellent for our life support and fuel needs.

All of the major moons and most of the minor moons of all the gas giants are tidal locked, just like our Moon, so their own rotation and orbit around Jupiter are the same length of time. This gives a major advantage we can make use of. Mounting a SE in the center of the "back" side gives us a ribbon that sticks straight out away from both the moon and Jupiter! Taking Ganymede as our example, the ribbon rotates with the moon around its own axis, just as an earth ribbon would do, at the rate of 7.16 earth days per revolution. In that same time the entire assembly, moon and ribbon, also completes one revolution around Jupiter. So, the ribbon is always orientated as if it were an extension of a radius from the center of Jupiter, through the center of the moon and continuing on. It gives us a swing arm distance, as if the ribbon ran all the way from the center of Jupiter!

What we get out of space elevators is an angular velocity, acting through a swing arm length, that translates into a linear velocity vector, when we release from the ribbon.

$$V_c = \omega R_c$$

That is; linear velocity (V_c) at a ribbon length position, equals, angular velocity (ω), in radians per second, times arm length (R_c). In the case of tidal locked bodies, that angular velocity is the same as the moon it is mounted on, all the way up the ribbon, but the swing arm length is from the center of the primary, Jupiter. So, we can have useful ribbons on bodies with slow turning rates that would otherwise make for unworkably long ribbons if they were by themselves.

The angular velocity (ω)of Ganymede, is 1.02e-5 rad/s, both for its turning rate and its orbital rate. The escape velocity (Ve) from Jupiter at Ganymede's orbital distance is 15.39km/s, while Ganymede's orbital speed is 10,870 meters/s, which is also the linear speed of the base of our ribbon.   (We will ignore Ganymede's, gravitational influence for right now in order to show just the Jupiter influence. Ganymede's gravitational factor will dwindle to insignificance anyway.)   As we move out on the ribbon we get further from Jupiter and this escape value drops lower while the linear velocity of a spot on the ribbon increases. All we have to do is find where the two values cross. That happens at 279,000km altitude on the ribbon when the Ve from Jupiter is 13.704km/s and the linear speed on the ribbon is 13.710km/s. Below this altitude we enter various ellipses inside the Jovian system, above this altitude we would escape to other places.

Now that we have shown that we can have a viable transportation system into and out of the Jovian system let's apply this to our scoop-ship. The idea would be to run the ship up the ribbon a short distance, just enough to offset Ganymede's gravity, accelerate the ship to skim past Ganymede (a small gravity assist) to fall toward Jupiter, in a long flat ellipse. We arrive at Jupiter at about 60km/s and skim into the upper atmosphere to scoop up our load. As we are scooping up mass we are losing speed so we fire off our engines to regain some of that speed and in doing so we burn off a fuel mass about equal to the mass we are ingesting. When we exit the atmosphere we have about the same mass that we started with and we have lost speed then added some back with engines, thus coasting to the vicinity of our starting point.

An alternate scheme is to draw all the fuel to burn, and more for the exit burn out side Jupiter, from the atmosphere, in addition to the amount being captured and tanked. Let's say that we have slowed to 43km/s which is orbital speed at the edge of the atmosphere, a convenient target number to manage our energy budget and not make our job any harder than necessary. We need to get back to the speed of the bottom of an ellipse that goes back to Ganymede's orbit or 58.9km/s so we need 16km/s from the engines exiting Jupiter. We will have to have stout engines but it is doable. At the end of the return trip, close is good enough, maneuvering engines and Ganymede's own gravity can assist us in coming around to a ribbon pickup. Of the tons of Jovian gas that we brought back only a small fraction is ^{3}He, virtually all of the load can be separated and re-tanked as fuel for the next trip, making up any difference from local supplies.

We have one other problem we must attend to. Above, we computed that at 279,000km on a Ganymede ribbon we have enough linear velocity to overcome Jupiter's escape velocity of 13.7km/s at that point. But for transport to Earth via minimum Hohmann orbit we need at least 14.481km/s for the proper insertion velocity and that happens at 354,765km ribbon altitude. (See Addendum for the math.) Going even higher on the ribbon, say 700,000km will give us a faster trip home with a very nice 4.76km/s in excess of the Hohmann minimum. The question is, how much ribbon can we use before running into Callisto? The distance between Ganymede and Callisto is 814,000km, so obviously that is our absolute limit. But our practical limit is governed by Callisto's gravitational effect, specifically any pumping action, that would be applied to the ribbon as the two moons pass each other every 12.53 days. As a first approximation we see that if we used a 700,000km ribbon and the end of our ribbon passed within 114,000km of Callisto the gravity gradient would only be 0.000566g. This is a totally insignificant force compared to those involved in the ribbon itself and would only become a problem if the small perturbations were reinforced each time around. These forces are small enough to be easily countered by modest thrusters mounted on the end of the ribbon, so it looks like we can plan on using at least 700,000km of ribbon.

For those who like large engineering projects you might want to work on this idea. It is theoretically possible to use one of the close-in moons of Jupiter and run down a ribbon and dip directly into the atmosphere. Metis is only 57,000km above Jupiter's "surface" and is 40km in diameter. It must be a captured stony or metal asteroid else it would not have escaped being broken up that deep in Jupiter's gravity well. Since its rotation is tidal locked we can attach a ribbon in the center of the side facing "down" and it isn't going to move relative to Jupiter. It would have to be a good-sized ribbon, for a lift capacity of 100 tons it would mass about 1 billion tons. Now, one might well ask, if we mount a billion ton ribbon and drag the end through the 400km per hour winds of Jupiter, wouldn't this tend to bother the moon a bit, drag it down? Well it might, but even though it is a very small moon it still masses about 95 trillion tons. Our ribbon would only be 0.0011% of that, so we ought to be able to get away with it for some time.

Actually, our more immediate problem would be to see if we could build a ribbon and climber machinery that would last in the heavy radiation fields. At this point we don't know what type of coating, or even if we could apply such coating, that would sustain the ribbon in those fields. Certainly our base of operations and handling facilities would have to be underground. While you are there, why not just drill a hole straight through the center of Metis and mount the operations base and ribbon anchor in the exact center of the moon? It is only 40km across and we have already dug tunnels that long here on Earth. The ribbon would hang down the center of the hole and straight down to Jupiter. The cargo can be off-loaded in shielded conditions at the hub and the refined products shipped up the hole to the "back" side of the moon where the shipping operations would enjoy some shielding. Shuttle ships would operate between Metis and Ganymede where cargoes would be consolidated on larger ships for ribbon launch to the inner system.

If Jupiter's heavy and harsh conditions proved too much to handle (not forever, just several more centuries to conquer) there are still the moons with plenty of resources to sustain permanent bases. Altogether the Jovian moons could support a considerable human presence, supplying the inner system with volatiles, and as way station to the planets yet further out. Ribbon launched transport can profitably work in both directions.

Saturn

Saturn, like Jupiter with its cadre of moons, is another mini solar system. At 9.54 AU it is almost twice as far out and more than twice the travel time, we will have to have large ships for generous living space to work this system. Saturn also has serious radiation problems to be overcome, and worse yet the extensive ring system pretty much precludes any ribbon operations from low moons dipping into Saturn. With an escape velocity of 35.2km/s, much reduced from Jupiter's, we may obtain scoop ship capability much sooner at Saturn. Scoop ships could operate out of the equatorial plane of the rings, but at close approach you do have to cross the plane of the rings, in order to get into and out of the atmosphere. That gap that we can see between the inner rings and Saturn may not be entirely free and clear of debris, we will have to go there and explore in detail to find out.

Saturn may be a hazardous hunting ground for ^{3}He but the consolation prize is Titan, at 5200km wide, only slightly smaller than Ganymede, (both of which are larger than the planet Mercury). Titan has a thick atmosphere, oceans, hard ice for land, and volatiles in abundance, everything we need for an extensive human habitat. The atmosphere is 1.5 Earth pressure and 90% nitrogen, an excellent source of the buffer gas we need for breathable atmospheres in all the habitats and ships we are developing around the solar system. Titan has an ethane (C_2H_6, from CH_4) ocean and is thought to have an abundance of Ammonia, (NH_3), water ice, methane and even a good supply of argon to lend a noble air. From all the hydrocarbons we could make "oil" and ship it to the rest of the solar system. Shipping would be expensive with present systems but it is a huge source, much larger than Earth's original unused fossil fuel supplies. Eventually we will tap it. At a minus 172 C Titan is much too cold for exposed skin but we would not need "space suits" just heated clothing and sealed breathing systems. The same applies to habitat construction, which could be much less costly and more expansive than for space travel enclosures. This makes for the best working and building conditions outside of Earth or a soon to be terraformed Mars.

With all this going for it, it's a shame that the main draw back, from Space Elevator Incorporated's point of view, is that it is not a good platform for an SE. (But it would be great for our inflatable habitat materials!) Titan revolves once around itself and around Saturn in 15.945 days, at an orbital distance of 1,221,800km. While the swing arm length is long, the slow rotation still requires a ribbon of 622,470km to get the 8.411km/s velocity needed for just the minimum Hohmann return trajectory. And that is way too long with the moon Hyperion only 259,300km out from Titan.

But there is an easy and quite unique solution. With an atmospheric pressure of 1.5 Earth normal and the rather chilly −170 C, the "air" density is 4.5 times Earth normal. We can build space ships as depicted in the old Flash Gordon comic books, with wings, and fly into orbit! A gravity of about 1/7 g and air density of 4.5 gives a lift to weight ratio, for a given area of wing, 31.5 times better than on Earth, so the wings don't have to be very large, in fact a lifting body shape might do it. A nuclear-thermal rocket can tank up on nitrogen or methane for the fuel it uses in space but ingest the local atmosphere as fuel during atmospheric flight, keeping the mass ratio very modest. We only need a ΔV of 1.18km/s to get to a 50km high parking orbit - piece of cake. We can probably do that while still in the atmosphere, giving us a free ride to orbit as far as on board fuel is concerned, the same condition as using a ribbon and it's faster. Our return from a dip into Saturn is equally economical since we can use aero-breaking and then an airplane landing. Round trip, we may actually use less energy than if we had a space elevator!

This idyllic situation still leaves one problem. Without a ribbon how do we get cheap transportation to the rest of the solar system? This is a classic example of "you can't get there from here you have to go some place else and start." Since we can get on and off Titan easily enough, we just go next door to moon Rhea to set up our space elevator. Rhea is 694,800km closer to Saturn giving us plenty of room to swing a cat. It rotates in 4.518 days giving us 3.5 times more angular velocity and it is small at 1530km for an easy to manage gravity of 0.07g. The swing arm is 527,000km from Saturn, (before adding ribbon length) and the angular velocity is 1.6e-5 rad/s. This will give us a Hohmann departure for Earth at a reasonably short, compared to what we have been working with,

194,234km ribbon length. At 400,000km we get an additional 4.47km/s ΔV for a speedy trip of 4 years to Earth, compared to 6.05 years for the regular Hohmann orbit. This last is just by way of showing that it could be done, we will leave it to others to figure out how to stop when we get to Earth.

Uranus

Do you remember the story of the three bears? The gist of it was that one was too much (of whatever) and one was too little and the third was just right. Well, Uranus may not be "just right" but at least it is a lot better than Jupiter and Saturn on radiation, heavy gravity and debris in orbit. Its main disadvantage is that it is so far out there, 19.2 AU, 16 years Hohmann orbit travel time; that's not a trip, it's a career. Fair to say that Uranus would not be developed until permanent posts were established in the Jovian and Saturn systems, and could be supported from there (when the planets are aligned). Cargo ships would be robotic and ribbon-launched into Hohmann orbits. We would just have to be satisfied with waiting to fill up the pipeline with a string of shipments. People ships would need to be of the fusion type able to double and triple the ΔV's obtained from the various ribbon launchers that we would now have around the other planets from Earth to Uranus. This would shorten the commute, at least from Saturn, to one to two years. Uranus is 50,100km wide, an escape velocity of 22km/s, and has a nice system of moons with lots of ice and volatiles to support a permanent station.

One of those moons, Miranda, seems to be mostly ice, in fact it looks like it was hastily assembled from gigantic blocks of ice; most untidy. Miranda is 480km in diameter, mass 66 quadrillion tons, (half of that ice) small enough to give virtually no gravitational hindrance to flight operations and just enough (>1%g) to give some gravity to living conditions. Not only is the ice and size good for a human habitat but also it is in just the right place for working Uranus. It orbits at 129,400km, or 104,400 above the surface and has a period of 1.414 days, fast enough to make a good ribbon platform.   We could work either scoop ships or ribbon dipping from here.

The problems involved here are, Uranus has a thin plane of debris rings, and the five major moons are close together. The moon next door is Ariel at 191,000km orbit, leaving only 61,600km of room for a ribbon. Both of these problems may be overcome by Miranda being 4.2 degrees out of the equatorial plane. This means that the dipping ribbon going down will go beside the rings most of the time, crossing through the rings twice each revolution. The engineering challenge will be to design something to survive this; at least it is better than being *in* the ring plane. As for the 'up" ribbon extending away from Uranus, at 61,600km a ribbon would swing above Ariel's orbit by 4,514km. Ariel's radius is 579km leaving 3,935km clearance – piece of cake. This would most likely require an end weight with motorized steering capacity to make darn sure of clearance when the orbits cross each others plane, and to take out any oscillations induced by such close passage. The next moon out is Umbriel, 137,000km away, and also in the plane so the ribbon would miss by 10,000km, most of the time. For a Hohmann escape orbit we need a ribbon length of 50,734km and for an extra 6.4km/s ΔV it would be nice to go out to 150,000km, sweeping past two of the outer moons. That would be a rather exciting display for passengers, with closing speeds of over 4.3km/s; quite literally, faster than a speeding bullet. Not counting, for the moment, the huge distance, and thus time, involved in dealing with Uranus, the mining of its ^{3}He would cost less energy, less engineering, and less facilities, in short, it would be more efficient than operations on either Saturn or Jupiter. Except for one little problem.

Uranus has chosen as its lot in life to rotate on its side, its axis tilted 98 degrees to the ecliptic, its north pole points, almost, at the Sun during its "summer" and the south pole during its "winter". This puts Uranus's equatorial plane, and the orbits of its moons, perpendicular to the plane of the solar system. Ships launched by centripetal force from the equatorial plane would not be heading for the other planets, they would be going perpendicular to the Suns ecliptic. "Turn right at the second moon." They would have to perform a huge and energy expensive plane change of about 90 degrees. Seven of the other planets have more modest tilts of less than 29 degrees so by this time we would be used to making plane changes after getting most of the work done by the ribbon launch. But a plane change of up to 98 degrees would leave to the ribbon only the work of getting out of Uranus's gravity, needing large engines and fuel requirements for the rest of the trip. There are a couple of ways to manage this less than optimum situation.

Twice in a Uranus year the edge of the equatorial plane points at the Sun, the poles pointing along the orbital path. Launches during these times can put all of the energy directly into the ecliptic plane and be very efficient for both arriving and departing traffic. The trouble is that the Uranus year is 84 Earth years long so it would be a long wait for these favorable conditions. We can lengthen the window of good ribbon launch conditions by many Earth years by accepting some small amount of plane change, which we would be using anyway in aiming at the various inner planets as they change position in their orbits relative to the position of Uranus. If we accept 30 degrees as a maximum plane change, that is 30 degrees either side of dead on to the Sun, then we will have a ribbon launch season of 14 Earth years, coming twice in a Uranus year. We can ship an awful lot of stuff in 14 years. What do we do for the 28 years we are waiting for the next season?

For the off-tourist season we can launch, more or less, along the orbital path of Uranus. That is still in the orbital plane, it's just not headed "down system" the way we would like. The ribbon can launch us back along Uranus's orbital path killing the 6.797km/s orbital speed, leaving us with zero speed relative to the Sun. We only have to go out just far enough to get away from Uranus's gravity influence and we can fall all the way to the inner system. For cargo this would be the ultimate in efficiency, requiring only steering and maneuvering engines/fuel. But this falling, from 2.87 billion kilometers, can take a long time to get under way, at any noticeable speed, so a little squirt from the engines would be most appreciated. As mentioned above we would already have very powerful engines by this time so for faster trips we use as much thrust as appropriate. The requirements are not large. This method of coasting out to a dead stop relative to the Sun and then applying 5 to 10km/s ΔV towards our destination is very efficient.   One can even argue that since the plane change energy equation has a velocity factor, and we are starting from zero when we turn the ship to point down system, we have eliminated the "plane change problem".

By the 23rd century we may have such automated systems that most, if not the entire project, from sending the first equipment to the return of product can be handled by remote control, from some comfy location like the beach at Olympus Mons.

What about Neptune? Well, you get the idea; we can use it too. Neptune has the last of the giant moons, Triton, with islands of frozen methane in an ocean of liquid nitrogen. We might even go out there just to tank up on these items when we get to the point of needing them in billion ton lots. We will get out there eventually, but really, we should save something for the next millennium to work on. How about a giant, free standing ribbon, one millionkm long, spinning about its center point for slinging starships to the next star system? You work it out.

Summary

Since its conception, the space elevator has appeared repeatedly in science fiction and concept overviews, it's an idea that just won't go away. With this manuscript we have attempted to present a complete quantitative analysis of the problems that will be encountered designing, constructing, deploying and utilizing a space elevator. We have started with the basic concept and existing technologies and put forth a system design and deployment scenario. We have addressed the environmental conditions this system must survive. We have presented one possible deployment schedule and a cost breakdown based primarily on existing systems. We have presented the design tradeoffs and offered guidelines for how they might be decided. We have explored the economics and how the space elevator could be utilized. What we have found is that an extremely valuable space elevator can be built in the near future with acceptable risk. The funding, while certainly substantial, is far smaller than the expected return – a first in space development projects.

This manuscript is not intended to be the final word on any aspect of the space elevator. It is intended to be a good beginning. The next stage is not to form large design committees but to get armies of graduate students and researchers examining each individual aspect of the scenario we have presented. Carbon nanotube composites, meteor impacts, weather, orbital mechanics of the deployment, induced oscillations from every source, power beaming, on-orbit operations, electric propulsion, atomic oxygen, nanotube and epoxy coating, climber design, ribbon spooling and ribbon design all need to be studied and can be done by many small programs. Once the individual efforts have produced the needed information, a technically-driven, fiscally-responsible team of committed individuals can be formed to build the space elevator.

One question that has been raised is where will the funding come from to build the first space elevator. The proposed project cost of $6 to $10 billion is not a large sum of money in today's world. There are many individuals, institutions and governments that invest such sums on a daily basis, too many for those nervous about who might build and control such a valuable system. The question is not whether funding for the elevator is available but who will jump at the chance first. With a concerted and objective effort we could begin construction of the space elevator in the coming decade.

We have looked into the future, as best we can, and have seen that to maintain a prosperous Earth and a prosperous future for Mankind, we must develop space. Beyond the research, beyond the exploration, one way or another, we must have a commercial, inhabited development of space. As the world's technology leader the US can maintain its present slow expensive way and try to do it before we run out of resources or we can strike out on a faster, cheaper, more productive way with space elevator technology.

The steam engine was the first machine to change Man's productivity by several orders of magnitude. Then electricity did it again, and recently the transistor has moved the decimal point another notch. The space elevator is awaiting its turn. This is the century that we will either become a space-faring society or a horse drawn society. If we succeed we will have ten's of thousands of years of prosperity, but even more important than that it will mean that one day we will send our children to the stars, and in that we will have our immortality.

"For I dipt into the future, far as human eye could see,
Saw the Vision of the world, and all the wonder that would be;

Saw the heavens fill with commerce, argosies of magic sails,
Pilots of the purple twilight dropping down with costly bales"
Locksley Hall

Addendum

For those of you who like higher mathematics, here is the integral expression of the space elevator. Enjoy. For a more practical approach, see the spreadsheet below.

From Pearson (1975) we have the expression:

$$A(r) = A_s e^{\left(\frac{3r_o^2}{2hr_s}\right)} e^{\left(-\frac{r_o}{h}\right)\left(\frac{r_o}{r} + \frac{r_o r^2}{2r_s^3}\right)}$$

for the cross-sectional area of the ribbon as a function of the orbital radius. From Pearson we also get:

$$\frac{A_s}{A_o} = e^{\left(\frac{r_o}{h}\right)} e^{\left(1 + \frac{r_o^3}{2r_s^3} - \frac{3r_o}{2r_s}\right)}$$

where:

$$h = \frac{\sigma}{\rho g_o} = \frac{\sigma r_o^2}{\rho M_p G}$$

σ = ribbon stress (tension/A(r))

ρ = ribbon density

g_o = the gravity at the planet's surface, $\dfrac{M_p G}{r_o^2}$

$$A_o = \frac{m g_o}{\sigma} = \frac{m M_p G}{\sigma r_o^2}$$

$$r_s^3 = \frac{M_p G}{w_p^2}$$

m as the mass of the load on the ribbon(with a safety margin),
M_p is the mass of the planet
G is the gravitational constant
r_o is the radius of the planet
r_s is the synchronous orbit radius
w_p is the angular velocity of the planet

Substituting in for A_o we get:

$$A(r) = A_o e^{\left(\frac{r_o}{h}\right)} e^{\left(1 + \frac{r_o^3}{2r_s^3} - \frac{3r_o}{2r_s}\right)} e^{\left(\frac{3r_o^2}{2hr_s}\right)} e^{\left(-\frac{r_o}{h}\right)\left(\frac{r_o}{r} + \frac{r_o r^2}{2r_s^3}\right)}$$

$$A(r) = A_o e^{\left(1 + \frac{r_o}{h} + \frac{3r_0^2}{2hr_s} - \frac{3r_0}{2r_s} - \frac{r_0^2}{hr} + \frac{r_0^3}{2r_s^3} - \frac{r_o^2 r^2}{2hr_s^3}\right)}$$

Then the mass of the ribbon is:

$$M_c = \int_{r_o}^{r_t} \rho A(r)\,dr = \int_{r_o}^{r_t} \rho A_o e^{\left(1 + \frac{r_o}{h} + \frac{3r_0^2}{2hr_s} - \frac{3r_0}{2r_s} - \frac{r_0^2}{hr} + \frac{r_0^3}{2r_s^3} - \frac{r_o^2 r^2}{2hr_s^3}\right)}\,dr$$

$$M_c = \rho A_o e^{\left(1 + \frac{r_o}{h} + \frac{3r_0^2}{2hr_s} - \frac{3r_0}{2r_s} + \frac{r_0^3}{2r_s^3}\right)} \int_{r_o}^{r_t} e^{-\left(\frac{r_0^2}{hr} + \frac{r_o^2 r^2}{2hr_s^3}\right)}\,dr$$

$$M_c = \rho \frac{mM_pG}{\sigma r_o^2} e^{\left\{1 + \frac{r_o}{\frac{\sigma r_o^2}{\rho M_p G}} + \frac{3r_0^2}{2\frac{\sigma r_o^2}{\rho M_p G}\sqrt[3]{\frac{M_p G}{w_p^2}}} - \frac{3r_0}{2\sqrt{\frac{M_p G}{w_p^2}}} + \frac{r_0^3}{2\frac{M_p G}{w_p^2}}\right\}} \int_{r_o}^{r_t} e^{-\left\{\frac{r_0^2}{\frac{\sigma r_o^2}{\rho M_p G}r} + \frac{r_o^2 r^2}{2\frac{\sigma r_o^2}{\rho M_p G}\frac{M_p G}{w_p^2}}\right\}}\,dr$$

$$M_c = \frac{\rho}{\sigma} \frac{mM_pG}{r_o^2} e^{\left(1 + \frac{\rho M_p G}{\sigma r_o} + \frac{3\rho}{2\sigma}\sqrt[3]{(M_p Gw_p)^2} - \frac{3r_0}{2}\sqrt[3]{\frac{w_p^2}{M_p G}} + \frac{r_0^3 w_p^2}{2M_p G}\right)} \int_{r_o}^{r_t} e^{-\frac{\rho}{\sigma}\left(\frac{M_p G}{r} + \frac{1}{2}w_p^2 r^2\right)}\,dr$$

Stepwise integration for the Space Elevator

A segment of the Ribbons vital statistics spreadsheet.

	A	B	C	D	E	F	G
	Safety Factor	CNT density	planet radius	Step size	Tensile	Lift (kg)	G
2	2	1300	6378000	0.01568	1.30E+11	900	6.67E-11
3						519.2579	PL
4						380.7421	CW
5		MG/r-rw2					
6	Select	Planet	Mass	Diameter	Period	Ang vel	Step
7	0	Venus	4.87E+24	12103	243	2.99E-07	10
8	1	Earth	5.98E+24	12756	1	7.29E-05	0.015679
9	0	Moon	7.35E+22	3476	27.31	2.66E-06	0.3
10	0	Mars	6.42E+23	6794	1.03	7.06E-05	0.02
11	0	Ganymede	1.48E+23	5262	7.16	1.02E-05	0.1
12	0	Callisto	1.08E+23	4800	16.69	4.36E-06	0.2
13	0	Titan	1.35E+23	5150	15.94	4.56E-06	0.2
14	0	Mimas	3.80E+19	398	0.94	7.74E-05	0.01
15	0	Model1	3.80E+19	398	0.47	0.000155	0.003
16	0	Model2	3.80E+19	398	0.24	0.000303	0.001
17	0	Model3	3.80E+19	398	1.88	3.87E-05	0.02
18		Selected values	5.98E+24	12756		7.3E-05	0.01568
22		Ribbon L = 106378km			CW/Ribbon ration = 73.3%		
23							
24	step size			Width at bottom			
25	100	9.772293		13.5309	40079.868	69468.15	29388.28
26	Radius, (km)	Acceleration (gravity – centripetal)	Tension	Ribbon cross area	Ribbon mass	Total launched mass	Counter-weight
27	6378	9.77E+00	8.80E+03	1.35E-07	0		
28	6478	9.47E+00	8.96E+03	1.38E-07	17.5901		
29	6578	9.18E+00	9.13E+03	1.40E-07	35.5188		
30	6678	8.91E+00	9.30E+03	1.43E-07	53.7818		
31	6778	8.65E+00	9.46E+03	1.46E-07	72.3754		
32	6878	8.40E+00	9.62E+03	1.48E-07	91.2953		
33	6978	8.16E+00	9.78E+03	1.50E-07	110.5378		
34	7078	7.92E+00	9.94E+03	1.53E-07	130.0986		
35	7178	7.70E+00	1.01E+04	1.55E-07	149.9741		

This spreadsheet computes the particulars for any size ribbon for any celestial object.

In this example the object selected is Earth, and in particular, it shows the examination of the starting conditions for the orbit launched initial ribbon and the first construction climber. By inputting the mass of the climber in F2 the total mass of the ribbon for the set conditions is shown in E25, along with its counterweight mass in G25.

Here is how it works.

First off, spreadsheets can be arranged in any number of ways, so there is nothing to say that this is the right way. In the top two rows we set down some of the constants we will be using. That the constant for gravity, G, fell in the G column is purely coincidental. We also just happened to use F2, 3 and 4 for our climber mass. F3 and F4 show the portions of the climber mass (that was input in F2) that will be, the payload of new ribbon and what will be added to the counterweight at the end of its trip. For working, transport climbers, such as 20 ton climbers, F3 and F4 are not relevant.

The rows 6 through 18 are used to set up the planetary factors that we need. This is a selection table. By putting a 1 beside Earth in column A, (and making sure that all other rows of column A have a zero) we cause the Earth's parameters to select out on row 18, which then becomes the source data for the other formulas of the columns. As you can see you can put any object of interest in this table and see what a proposed ribbon would look like. Even unrealistic ones. Not included in this display, because of space limitations, are columns off to the right that show some ribbon results, such as length. This would show, for instance, that a ribbon for the Moon doesn't work because for a decent load it would be so long that it would extend well past the Earth, not very practical. With this selection table we can explore the possibilities of using ribbons on various moons and even hypothetical objects such as manmade habitats and asteroids, (assuming that they rotate).

Row 25 is used to extract results from the table below, so that you don't have to keep paging down to find them. A25 notes the step interval of integration used for the length of the radius arm (the ribbon) from the center of the Earth. Values are in meters. Note that the starting value is the radius of the Earth, that is, the surface of the Earth and each step increases by 100,000 meters (100km). Depending on the memory available and the time one has to create thousands of rows for this table you can make this step interval as small as you want. At a step interval of 100km the estimated error is

0.4%, more than adequate for engineering studies. By the time you refine it any closer than that you are into the cumulative error rate of the set of constants. As for instance, gravity varies from place to place on the Earth, so you would have to go out and measure the actual gravity factor at the anchor site.

The table gives values for any length so you have to decide what length you want to look at. In this case, for a ribbon length of 100,000km we find the row for 106378000 meters and set cells E25, F25, and G25 to report the results of that row. Now you can change the size of the load in cell F2 and see the resulting total ribbon mass and counter weight on row 25, for that length. This table also shows the profile, the taper, of the ribbon. Column D gives the cross sectional area of the ribbon at each altitude, for the given load.

The columns for launch mass and counterweight are empty in this display because there are no meaningful values until you get out past GEO altitude.

The formulas for the columns are:

A27: C2, A28: A27+$\underline{D2}$*$\underline{C2}$

B27: $\underline{H2}$*$\underline{G2}$/A27^2 – A27*$\underline{I2}^2$

C27: F2*B27, C28: (C27+D27*(A28-A27)*(B27+B28)/2*$\underline{B2}$)

D27: C27*$\underline{C4}$/$\underline{E2}$

E27: 0, E28: D27*(A28-A27)*$\underline{B2}$

Note that the first cell (row 27) in column A, C and E are not the same as the cells below that. Underlined items are constant references that do not change in lower cells; (often denoted by a $ in spreadsheets).

Calculating Delta V's For
Interplanetary Orbit Insertions
By Edward Wullschleger

Introduction

There are some common errors that are often made when trying to calculate insertion velocities and delta V's for interplanetary orbits that begin in Earth orbit. The purpose of this article is to identify those errors and explain how to avoid them.

This explanation assumes that a vehicle is already in a circular Earth orbit, commonly referred to as LEO, and that we need to calculate the additional "delta V" needed to escape Earth and enter a Hohmann transfer orbit that will take us to the vicinity of another planet. This is the simplified version, a plane change maneuver, normally required for planetary missions, is not included here because of the variable starting conditions that could effect such a maneuver. We simply note that any real world trip will take more total energy than calculated by these equations.   It is assumed that the required starting and ending positions of the Earth and the other planet, for the respective departure and arrival times, have already been determined. Otherwise the proposed

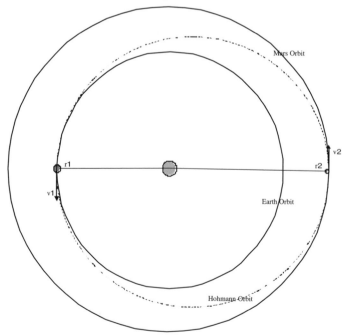

Figure 1 Hohmann Orbit Diagram

maneuver would still get you to the orbit of the other planet but without the planet. The essential focus will be on velocity values.

The problem is composed of two parts. The first part relates to the details of the heliocentric Hohmann transfer orbit. The second part refers to the hyperbolic Earth orbit required to enter the Hohmann transfer orbit.

The Hohmann Orbit

As shown in Figure 1, the Hohmann orbit is an elliptical orbit, with perihelion at Earth orbit and aphelion at the other planet's orbit (in this example, Mars). The Hohmann orbit is the minimum energy transfer ellipse between orbits; it is also the slowest. We could make faster trips by using more energy than these equations dictate.

The Hohmann orbit is described by its perihelion, r_1, the distance from the Sun to the Earth, and its aphelion, r_2, the distance from the Sun to destination planet, (Mars). The velocity V_{r1} is the starting velocity of the Hohmann orbit, at Earth orbit, with respect to the Sun and V_{r2} is the velocity at the destination (Mars), and can be calculated using the equation:

1) $V = (G*(M+m)*(2/r - 1/a))^{1/2}$ (Reference 1, p.49)

where G is the gravitational constant, M is the mass of the Sun, m is the mass of the vehicle, r is the distance from the Sun to the specified position on the orbit (r_1 or r_2), and a is the semi-major axis of the Hohmann orbit (H).

Unless you are driving around in an asteroid, the value of m is negligible compared to M, so we drop it and $G*(M+m)$ becomes $G*M$ and is calculated as follows:

$G = k^2$, where k is the Gaussian gravitational constant: 0.01720209895 (in AU's, Solar Masses, and Mean Solar Days), or 8.16822153915E-6 in mks units.

K(mks) = 8.16822153915E-6   (Reference 3, p.696, Datum 1.)
M = 1.9891E+30kg.   (Reference 3, p.697, Datum 19.)

So: $G*M = k^2 * M = 1.327124E+20 \ m^3/s^2$
        (Reference 3, p.696, Datum 16.)
(Since they are used together more often than not, the value $G*M$ is often referred to by the Greek letter mu and called "mu Sun". There is another mu for G times Earth Mass called mu Earth.)

And: a = (r1 + r2)/2   (Average of the two distances; e.g. a = 1.887673E+11)

The starting values for the orbits are:
r1 = 1.495979E+11 meters (Earth orbit, semi-major axis)
        (Reference 3, p.316, 1.00000011 * AU)
r2 = 2.279367E+11 meters (Mars orbit, semi-major axis)
        (Reference 3, p.316, 1.52366231 * AU
        with p. 696, Datum 12, AU = 1.49597870E+11)

Finally: The velocities in the H orbit, with respect to the Sun.
V_{r1} = 3.272933E+04 m/sec, (at r = r_1)   low point of H orbit, (at Earth orbit)
V_{r2} = 2.148096E+04 m/sec, (at r = r_2)   high point at Mars.
        (Note: these velocities are for the H orbit. V_{r1} while at Earth orbit is faster than Earth and V_{r2}, while at Mars orbit is slower than Mars.)

The Hyperbolic Earth Orbit
At this point it is very easy to think that we are almost done and simply subtract the Earth's orbital velocity around the Sun (2.978473E+04 m/sec), from V_{r1}, and believe that we have the insertion velocity needed relative to the Earth. However, this is not the case. This subtraction is a necessary step, but it gives us the "hyperbolic excess velocity" relative to the Earth; not the insertion velocity. The insertion velocity must be greater than the hyperbolic excess velocity because the vehicle must still escape from the Earth's gravitational field (Figure 2.)

2) $\Delta V_{r1} = V_{r1} - V_E$ 3.27283e4-2.97847e4 = 0.29436e4 = (2.944km/s)   (Ref. 2, p.364)

where: ΔV_{r1} is the hyperbolic excess velocity, V_{r1} is the Hohmann orbit velocity at the Earth's position, and V_E is the (average) velocity of the Earth in its orbit around the Sun (2.978473E+04 m/sec calculated using Eq.1 and r1). Another way to put the ΔV_{r1} velocity in perspective is to say that if you were traveling somewhere in the Earth's orbit, but without the Earth, this additional velocity would take you to Mars's orbit. The V_{r1} is the velocity needed to start an ellipse that swings from one planet's orbit to the other without staying at either orbit. But we do have Earth interfering with our start so we have to have an insertion velocity.

In order to calculate the insertion velocity, V_I, we need to account for the energy used to escape the Earth's gravitational field. Referring to Figure 2 again, we calculate the energy at two points relative to the

Earth. The first point is at the position in our orbit around the Earth just after we apply the delta V to achieve the insertion velocity, V_I. The second is at a point that we consider to be "at infinity" relative to the Earth, i.e., at a point where the Earth's gravitational field no longer has a significant effect on us (this is normally assumed to be somewhere around one million kilometers from the Earth. In equation 4, below, we call this point r_∞). (Reference 2, pp.368-369)

At the first point:

3) $E_1 = m(V_I^2)/2 - GmM_E / r_c$
          (1st term is Kinetic Energy, and 2nd term is gravitational Potential Energy)

where: V_I is the insertion velocity relative to the Earth after we apply the delta V, r_c is the distance from the center of the Earth to our position in Low Earth Orbit where the delta V is applied, m is the mass of the object (ship) and M_E is the mass of the Earth. (As in equation 1, GM_E is the now the value "mu" Earth, worth 3.986005E+14.) Therefore, $GmM_E = m * 3.986005E+14$ (Reference 3, p.696, Datum 6.) Now if you wanted to drop an object and find its gravitational Potential Energy, GmM_E is the equation to use, but we don't need that so, in a moment, we will find a way to get rid of the m factor.

At the second point:

4) $E_2 = m(\Delta V_{rl}^2)/2 - GmM_E / r_\infty$
          (Note that 2nd term, Potential Energy, equals zero due to division by r_∞)

Recall that Equation 3 is calculated just after the delta V burn, and Equation 4 is calculated after we have coasted to a point where we assume we have escaped from the Earth's gravitational field. Since no force other than the Earth's gravitational field has been applied between the two calculations, and since we account for that force in the Potential Energy terms, we can apply the conservation of energy law to say that $E_1 = E_2$

Therefore:

$m(V_I^2)/2 - GmM_E / r_c = m(\Delta V_{rl}^2)/2$

Dividing all three terms by m and multiplying by 2, we get: $V_I^2 - 2GM_E / r_c = \Delta V_{rl}^2$

And rearranging the equation gives the needed insertion velocity:
          Note using $r_c = 6378140 + 300,000 = 6,678,140$ m

5) $V_I = (\Delta V_{r1}^2 + 2GM_E / r_c)^{1/2}$ $((0.29436e4)^2 + 7.972e14/ r_c)^{1/2}$ $= 11{,}316\text{m/s}$

(this logic also in Reference 2, p.369, Eq. 8.3-10)

Note that the second term by itself, $(2GM_E / r_c)^{1/2}$, would be the escape velocity from the starting altitude r_c. Set r_c to 6378140m for escape velocity from ground level. Substitute the mass of some other body for M_E to get the escape velocity for that body.

V_I then is the velocity we need to obtain, relative to the Earth. Starting in orbit we already have some of that velocity. Our original circular orbit velocity can be calculated as:

6) $V_c = (GM_E / r_c)^{1/2}$ (where: $GM_E = 3.986005E+14$) $= 7726$ m/s
(low Earth orbit circular velocity at 300km altitude)

And delta V for the trip is simply: (for r =300 + 6378km)
The delta V that you need, V_I, minus the delta V that you already have V_c.

7) $dV = V_I - V_c$ $= 3590$ m/s

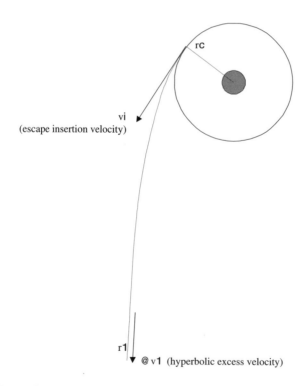

Figure 2 Earth Escape Diagram

References

1. Introductory Astronomy and Astrophysics; Smith and Jacobs. W.B. Saunders Company, 1973
2. Fundamentals of Astrodynamics; Bate, Mueller, and White. Dover Publications, Inc., 1971
3. Explanatory Supplement to the Astronomical Almanac; Seidelmann. U.S. Naval Observatory, 1992

[Continuing from where Ed left off...]

Dealing with smaller ellipses.

From Kepler's first law we know that all orbits are ellipses.

For working with ellipses that are all within the influence of one body such as orbits around Earth we can use a different version of the above equations. For V, the velocity of an object in an elliptical orbit we have what is commonly called the ellipse equation.

1) $V_e = (GM(^2/_R - ^1/((x+y)/_2)))^{1/2}$

Lets break this down to see how it works. Conditions: Earth with satellite.

We already know about G and M from above, in this case M is the mass of the Earth so we can shorten this portion to "mu Earth" or the constant 3.986005E+14, and we have done away with two of our terms already. Now 3.986×10^{14} is a very large number, almost a million billion of something, whereas velocities in space are in the thousands of meters per second. So we have reason to believe that what's on the right side of the expression will have to be a very small number to get us into the range of velocities that make sense.

R is the distance from the focus in use to any point on the ellipse one wishes to examine for the velocity of the satellite at that point. Or, in orbital terms, the distance from the center of the Earth to the point in question on the orbit; in meters. Usually, only two points are of interest, those on the major axis at the far end of the ellipse and the near end of the ellipse. (The Earth, or other primary object always occupies one of the foci of the ellipse, "the focus in use", the other is not used. For two large objects such as the Earth and the Moon, the primary object wobbles about its focus under the others gravitational influence. This makes real space travel in the Earth Moon system very complicated, mathematically, but we don't have to get *that* detailed.)

The x and y terms are the distance from the focus (in use), along the major axis to the far end of the ellipse, the apogee – set that to x – the other, y is then the perigee, the short way from the focus, in the opposite direction, along the major axis to the closer end of the ellipse. (You can set x or y to either one as long as you know what you mean.)   In terms of an orbit, that makes the apogee the high point of the orbit and the perigee the low point. The $(x+y)/2$ part of the expression is just averaging the apogee and perigee distances before setting it under 1 to make a fraction.   Once you have values for x and y, the whole $(2/_R - 1/((x+y)/_2))$ part of the equation simplifies to the form $(2/_R - 1/a)$.   If a were equal to R then the expression simplifies still further to $(2/_R - 1/_R)$ which collapses to $1/_R$ and our original equation becomes simply

4. $V = (GM/_R)^{1/2}$

When would a equal R? When x = y, which happens when the ellipse is a circle! Yes, the circle is a special case of the ellipse with an eccentricity of zero. So, for circular orbits about the earth, to get the velocity of the object just divide the constant mu by the distance from the center of the Earth and take the square root of that.

Now, getting back to the complex form of the equation, for elliptical orbits we usually want to know the speed of an object at the apogee and the perigee, so we set R equal to one of those and work the equation. Note that for these two points of interest R will be equal to either x or y.

Let's apply this to our ribbon where we are going to drop vehicles into low Earth orbits off the ribbon. We want to know how much ΔV we will need to change the ellipse we are in when we drop from the ribbon into a (near) circular orbit.

All the altitudes on the ribbon below GEO have a rotational speed lower than the speed that would be necessary to maintain a circular orbit at that altitude. So, when we disconnect from the ribbon we are going to drop into a lower orbit, an elliptical orbit with its apogee at the altitude we dropped from. Where will the low end of the ellipse end up? It might intersect the atmosphere or the Earth and end its travels. We need one more equation, the rotational speed where we dropped from the ribbon.

3) $V_c = 2R* \pi/period$ where the period is the sidereal day, 86164 seconds.

As an example let us say that we make the drop from 23,000km (or 29,378km from the center of the Earth.) Our ribbon speed is then: $V_c = 2.142$km/s.

If we wish to go into a circular orbit at that altitude then we work equation 2 and find that it takes a velocity of 3.684km/s. We subtract V_c from the circular velocity and see that it will take an additional 1.542km/s.   The idea is the same for entering the appropriate ellipse and then, later, changing that to a circular, usually lower, orbit. We already have x, the apogee at 23,000km and we should know the LEO altitude we want to get to, let's say that is 300km (6678km Earth center) which becomes y, the perigee. If we set R = x then by equation 1 we will get the V_e we need soon after departure from the ribbon in order for the low end of the ellipse to be 300km. $V_e = 2.242$km/s

By subtracting V_c from V_e we get 0.1km/s as the *additional* velocity, right off the ribbon, that we need to enter an ellipse with the low end at 300km altitude.

Just as the velocity in an ellipse at the apogee (high end) is slower than the corresponding circular orbit, (as we just showed) so also must the velocity at perigee (low end) be faster than a corresponding circular orbit. (Else it would not swing out again to return to the high point of the ellipse.) Therefore, having dropped in our ellipse to perigee, 300km, we must slow down at that point to circularize and stay in a 300km orbit. By how much do we need to slow down?

Compare the speed at the low point of the ellipse to the speed if it were a circular orbit at the same altitude; the difference is the ΔV we must apply to circularize the orbit. By equation 1, with R now set at perigee, 300km, (x & y remain the same, we are measuring the same ellipse at a different point, controlled by R) we get: $V_e = 9.864$km/s. From equation 2 we get V = 7.727km/s for a circular orbit at 300km.

Therefore, V - V_e = 7.727 - 9.864 = -2.137km/s
(A negative ΔV, be sure that you turned your ship around before firing!)

References

Angel, R. and Fugate, R.. 2000. *Science* 288:455.

Artsutanov, Y. 1960. V Kosmos na Elektrovoze, Komsomolskaya Pravda, (contents described in Lvov 1967 *Science* 158:946).

Asimov, I. 1972. *Biographical Encyclopedia of Science & Technology.* Avon Books

Asimov, I. 1975. *Jupiter.* ACE Books

Bennett, Harold. 2000. Compower, private communication.

Bouquet, F.L., Price, W.E., and Newell, D.M. 1979. Designer's guide to radiation effects on materials for use on Jupiter fly-bys and orbiters. *IEEETransactions on Nuclear Science* NS-26, no. 4:4660.

Brown, W.C., and Eves, E.E. 1992. Beamed Microwave Power Transmission and Its Application to Space. *IEEE Transactions on Microwave Theory and Techniques* 40, no. 6:1239.

Chelton D. B., Hussey K. J. and Parke, M. E. 1981. *Nature* 294:529.

Cheng, H.M., et.al. 1998. *Chemical Physics Letters* 289:602.

Choi, Y. C., et.al. 2000. *Applied Physics Letters* 76, no. 17: 2367.

Christian, H.J., Blakeslee, R. J., Boccippio, D. J., Boeck, W. L., Buechler, D. E., Driscoll, K. T., Goodman, S. J., Hall, J. M., Mach, D. A., and Stewart, M. F. 1999. Global Frequency and Distribution of Lightning as Observed by the Optical Transient Dector (OTD). Proceedings of the 11th International Conference on Atmospheric Electricity, Guntersville, Alabama. 726-729. Also see http://thunder.msfc.nasa.gov/data/

Clarke, A.C. 1978. *The Fountains of Paradise.* New York: Harcourt Brace Jovanovich.

Clarke, A.C. 1979. The Space Elevator: 'Thought Experiment', or Key to the Universe. *Adv. Earth Oriented Appl. Science Techn.* 1:39.

D'Amato, F.X., Berak, J.M., and Shuskus, A.J.. 1992. Fabrication and Test of an Efficient Photovoltaic Cell for Laser Optical Power Transmission. *IEEEPhotonics Technology Letters* 4, No. 3: 258.

Daly, E.J., Lemaire, J., Heynderickx, D., and Rogers, D.J. 1996. Problems with Models of the Radiation Belts. *IEEE Transactions on Nuclear Science* 43, no. 2:403.

Dissing, D., Verbyla, D. 1999. Landscape Interactions With Thunderstorms in Interior Alaska. In preparation. private communication.

Edwards, B. C. 2000. Design and Deployment of a Space Elevator. *Acta Astronautica.* In Print.

Egusa, S. 1990. Anisotropy of radiation-induced degradation in mechanical properties of fabric-reinforced polymer-matrix composites. *Journal of Materials Science* 25:1863.

Emberly, Paul. 2000. Kvaerner corp. Private communication.

Epstein, R. 2000. Private communication.

Forward, R. L. 1995. *Indistinguishable From Magic.* Baen/Simon & Schuster

Glaser, P.E. 1992. An Overview of the Solar Power Satellite Option. *IEEE Transactions on Microwave Theory and Techniques* 40, no. 6:1230.

Gribbin, John & Mary. *Fire on Earth.* Simon & Schuster

Heiken, G. 2000. Private communication.

Ho, C., et al. 1993. *Proc. SPIE* 1951:67.

Hoyt, R. 2000. NASA's Institute for Advanced Concepts Phase I Final Report.

Iijima, S. 1991. *Nature* 354:56.

The Infrared and Electro-Optical Systems Handbook: Atmospheric Propogation of Radiation, Vol. 2. 1993. Editor:Fredrick G. Smith. Bellingham Washington: SPIE Optical Engineering Press. 228.

Isaacs, J.D., Vine, A.C., Bradner, H., and Bachus, G.E. 1966. Satellite Elongation into a true 'Sky-Hook'. *Science* 151:682.

Koert, P., Cha, J.T. 1992. Millimeter-Wave Technology for Space Power Beaming. *IEEE Transactions on Microwave Theory and Techniques* 40, no. 6:1251.

Koert, P. 1999. APTI corp. private communication.

Kwon, Y. K., Lee, Y. H., Kim, S. G., Jund, P., Tomnek, D., and Smalley, R. E. 1997. Morphology and Stability of Growing Multi-Wall Carbon Nanotubes. *Phys. Rev. Lett.* 79:2065.

Lamontage, C. G., Manuelpillai, G.N., Taylor, E. A., and Tennyson, R.C. 1999. *International Journal of Impact Engineering* 23:519.

Lewis J. S. 1996. *Mining the Sky.* Addison-Wesley

Li, F., Cheng, H.M., Bai, S., Su, G., and Dresselhaus, M. 2000, *Applied Physics Letters,* 77, no. 20, in print.

Lipinski, R. J., et.al. 1994. *SPIE:Laser Power Beaming* 2121:222.

Loftus, J. P., and Stansbery, E. G. 1993. *Protection of Space Assets by Collision Avoidance,* IAA 6.4-93-752, 44th Congress of the International Astronautical Federation, Austria.

Manning, L.A., and Eshleman, U.R. 1959. Meteors in the Ionosphere, *Proc. IRE* 47:191.

Moravec, H. P., 1977, <u>A Non-Synchronous Orbital Skyhook</u>, 23rd *AIAA Meeting, The Industrialization of Space,* San Francisco, Ca., October 18-20, also *Journal of the Astronautical Sciences* 25, October-December.

Nagle, R. K., Saff, E.B. 1996. *Fundamentals of Differential Equations and Boundary Value Problems,* Second Edition. Addison-Wesley Publishing. 580.

Pearson, J. 1975. The Orbital tower: a spacecraft launcher using the Earth's rotational energy. *Acta Astronautica* 2:785.

Pournelle, J. 1979. *The Endless Frontier.* ACE Books

Pournelle, J. 1979. *A step Farther Out.* ACE Books

Ren, Z. F., et al. 1998. *Science* 282:1105.

Rohweller, D. 1999. TRW-Astro Aerospace, private communication.

Sandwell D. T. and Agreen R. W. 1984. *Journal of Geophysical Research* 89:2041.

Smitherman Jr., D. V. 2000. *Space Elevators: An Advanced Earth-Space Infrastructure for the New Millenium*, NASA/CP-2000-210429.

Space Mission Analysis and Design. 1991. Editors: J.R. Wertz and W.J. Larson, Kluwer Academic Publishers.

Stanley-Robinson, K. 1993. *Red Mars*. Bantam Spectra.

Staubach, P., Grün, E., and Jehn, R., 1997. *Adv. Space Res.* 19, no. 2:301-308.

Taylor, E. A., Herbert, M. K., Vaughan, B. A. M., and Mcdonnell, J. A. M. 1999. *International Journal of Impact Engineering* 23:883.

Yakobson, B.I. and Smalley, R.E. 1997. Fullerene nanotubes: C1,000,000 and beyond. *American Scientist* 85:324.

Yu, M. F., Lourie, O., Dyer, M. J., Moloni, K., Kelly, T. F., and Ruoff, R.S. 2000a. *Science* 287:637.

Yu, M. F., Files, B. S., Arepalli, S., and Ruoff, R.S. 2000b. *Physical Review Letters* 84, no. 24:5552.

Zhu, H. W., Xu, C. L., Wu, D. H., Wei, B. Q., Vajtai, R., Ajayan, P. M., 2002. Direct Synthesis of Long Single-Walled Carbon Nanotube Strands, *Science*, **296**.

Zubrin, R. 1996. *The Case for Mars.* Free Press.

Zubrin, R. 1999. *Entering Space.* Tarcher/Putnam.

Index

D

E

T